Essentials of Civil Engineering Materials

First Edition

By Kathryn E. Schulte Grahame, Steven W. Cranford,

Craig M. Shillaber, and Matthew J. Eckelman

Northeastern University

SAN DIEGO

Bassim Hamadeh, CEO and Publisher

John Remington, Executive Editor

Gem Rabanera, Project Editor

Abbey Hastings, Associate Production Editor

Emely Villavicencio, Senior Graphic Designer

Trey Soto, Licensing Coordinator

Natalie Piccotti, Director of Marketing

Kassie Graves, Vice President of Editorial

Jamie Giganti, Director of Academic Publishing

cognella® | ACADEMIC PUBLISHING

3970 Sorrento Valley Blvd., Ste. 500, San Diego, CA 92121

Detailed Contents

Introduction to Engineering Materials

I. Why Do Civil Engineers Need to Study Materials?

The simple answer to the aforementioned question is that civil engineers *build* things—and building anything requires *materials*. Everything—built or natural—is made up of materials of some kind. Try to envision a bridge built *without* materials. That doesn't even make sense! Clearly, materials are important.

Answered another way, if we as civil engineers are to be *trusted* to build things, we want to reliably predict how the materials in those things (whether a building, a bridge, or a roadway) behave under the conditions for which they were designed. However, we, as good practicing engineers, want to *smartly* choose the materials we use, which requires some basic knowledge of the *science* of materials. Clearly, all disciplines of engineering, from civil to mechanical to electrical, need a working knowledge of materials. There is also an underlying motivation (but not necessarily a requirement) to select the *best* quality and type of materials for a given project's specifications, not simply materials that "just work."

The highly interdisciplinary field of **materials science** involves the discovery and design of *new* materials, as well as the increased understanding and characterization of *all* materials. Critical to societal needs, materials science can be viewed as a direct complement to technological progress—aside from the Romans' use of concrete, classes of materials have been used to classify stages of civilizations, ranging from the Stone Age to the Bronze Age, and the current Silicon Age (or perhaps Semiconductor Age). Many of the most pressing scientific challenges humans currently face are due to the *limitations* of the materials that are available and, as a result, breakthroughs in materials science are likely to have a significant effect on the future of technology. What are some limitations of current materials? Think about it.

At the same time, there is also a trend to explore and exploit smaller and smaller material scales as modern scientists use nanotechnologies to enable analysis and design at the molecular level (i.e., from the "bottom up"). In nanoscience, researchers use the latest methods in applied physics and chemistry to quantify the relationship between a material's atomic and molecular structure and the emergent engineering properties. No longer is knowledge of macroscale properties sufficient—one must also be familiar with atomistic and molecular features (e.g., a basic knowledge of chemistry and physics). This has led to advances in metal processing—for example, manipulating the molecular structure to produce higher-strength steels for construction.

Civil engineering, in particular, is the engineering discipline that deals with the design, construction, and maintenance of the physical and naturally built environment, including works such as roads, bridges, canals, dams, and buildings. As a result, our concern, from a materials perspective, is the *building materials* of those works, such as concrete, steel, and aggregates. That being said, from a civil engineering perspective, there seems to be an overreliance on well-used and well-trusted materials, such as steel and concrete. These materials have formed the basis of urban infrastructure for over one hundred years, and engineers in particular like things that "just work." But just as cutting-edge plastics and alloys have replaced less-advanced materials in commercial goods, there is an increasing push to replace steel and concrete with high-performance "designer" materials that are stronger, tougher, cheaper, greener, and more efficient. Perhaps the next super-material for construction is in the near future.

II. Basic Requirements of Engineering Materials

While today's civil engineers need not be material specialists, a basic understanding of a material's *selection* and subsequent *performance* are essential to design, construction, and sustainment. For example, proper materials selection is an essential link between design and structural mechanics. In structural mechanics, we analyze loads, internal stresses, and deformation of members, and in design, we select the member materials and the ultimate shape of a structure to carry those loads.

When considering if a material is suitable for engineering purposes (whether civil, mechanical, electrical, etc.), several requirements are to be met. Engineering materials *must* be as follows:

1. **Understood**: How the material behaves when put to use must be known. From experience, you probably wouldn't build a road out of cotton balls. Without knowing the technical properties, you know that a stronger material is required. Before materials were rigorously characterized and categorized, tradesmen knew how certain materials performed by experience. We must understand material behaviors, such as the strength when subjected to load, if it stretches like a rubber band or if it is stiff like a rock, if it weakens when heated or if it gets brittle when cold.

2. **Reliable:** The material must have consistent properties and be durable enough to last the intended lifetime of its application. For example, it would be difficult to construct a house made of wood if the strength was unpredictable. Likewise, consideration must be taken if the materials degrade and/or corrode and lose strength.

3. **Workable:** The material must be easily shaped, handled, fastened, connected, and so on to enable construction of various components and structures. One of the reasons steel is so desirable as a construction material is its ability to be formed into different structural shapes and easily connected (via bolts or welds). Likewise, concrete can be cast into many different forms. A material like diamond, for example, is extremely strong, but would be almost impossible to form into anything other than small stones. Some metals, such as titanium alloys, are known as difficult-to-machine materials—many machine shops avoid titanium because of its reputation as a tool killer that is unreasonably expensive to machine.

If the material is well understood, reliable, and workable, we *can* build something from it. Moreover, we can likely predict how the final product will behave. However, those are not the only requirements we impose on our material selection. Good engineers strive to impose **efficiency** in their designs. Efficiency can mean a lot of things, but usually there is a balance of strength (to carry a load) as well as self-weight (which adds to a load) and geometric constraints. The ancient Egyptians, for example, were able to build the pyramids out of stone, but the base of the pyramids had to be extremely large compared to their height in order to carry the weight. The material (relatively weak limestone) is not very efficient to construct large buildings—but it served well at the time because the Egyptian craftsmen understood the properties of limestone, and it was workable. Modern buildings clearly do not use limestone. Efficiency can also imply durability (long lasting), ease of construction (connections), meeting time constraints (curing times), and, of perhaps most importance, cost.

Related to cost (supply and demand), the final general requirement engineers must consider for material selection is **availability**. Sometimes the best material for the job is simply not convenient (or too costly) to acquire—thus, consideration must be made for alternatives that are easily attained. For example, while steel is relatively easy to acquire in urban areas, more rural areas may benefit from cast-in-place concrete. Globalization of materials supply has somewhat alleviated this problem. Availability is one of the reasons steel is so predominant as a construction material. Steel is produced by iron ore,[1] which

---

1   There is more discussion in **chapter 5** on the chemical composition of steel.

happens to be one of the most available raw resources. Around 98 percent of the mined iron ore is used to make steel. In terms of importance to the global economy, only oil is comparable to iron ore for substances we mine from the ground. Estimates put available and accessible iron ore deposits on the order of 20 billion metric ton. If we preferred to build structures out of platinum, for example, there is only approximately 50,000 metric tons on the entire planet—about one-tenth the weight of the Burj Khalifa.

Beyond the general requirements discussed earlier (understood, reliable, workable, efficient, available), and alongside simple function (will it work?), when selecting a material, civil engineers have other socioeconomic factors to consider:

1. The primary criterion should always be **safety**. As both the National Society of Professional Engineers and the American Society of Civil Engineers **Code of Ethics** dictate,

   Engineers, in the fulfillment of their professional duties, shall hold paramount the safety, health and welfare of the public.[2]

   Whether selecting steel for a skyscraper or a recycled glass concrete for a walkway, always know that even if safety is not specifically written in the design criteria, it is most definitely implied for all materials that are to be selected.

2. **Economics** is another important design criterion (and likely the most restrictive to any large design or build project). Every project will have a budget that limits how materials are selected. Since multiple materials may share the properties specified, it is up to the engineer to select the most appropriate material for each project's budget. However, the cost of materials has and will continue to change with time. Unforeseen events, such as a natural disaster, might result in production issues of certain materials that suddenly raise their cost or completely limit their availability. Construction issues may also play a part in material selection. Timing is critical on many projects; therefore, materials with faster delivery or curing times may trump an initial decision to choose a less costly material.

3. **Sustainability** is a more recent addition to engineering design. From one perspective, sustainability refers to the overall impact a material's use will have both on the present and future environment. We—as civil engineers—want to ensure that future generations can rely on the continued use of a material supply indefinitely. Sustainable development and engineering consists of balancing local and global

---

2   National Society of Professional Engineers (NSPE), "Code of Ethics," https://www.nspe.org/resources/ethics/code-ethics; American Society of Civil Engineers (ASCE), "Code of Ethics," July 29, 2017, https://www.asce.org/code-of-ethics/.

efforts to meet basic human needs and build infrastructure without destroying or degrading the natural environment.

Regardless of the material system or intended use, these considerations are implicit for any engineering application and/or design. First and foremost, good knowledge of material *properties* is necessary to even compare materials for possible selection, which is the topic of the next section.

III. Material Properties

Material properties are the fundamental descriptors of a material. When comparing two materials (be they steel and concrete, straw and sticks, or feathers and bricks), we can list characteristics to describe how the material looks or behaves, or some features of the material that can be objectively measured, to differentiate between the two. These are the material *properties*. For engineers, there are some properties that are critical for technological applications (such as strength or thermal conductivity), but others are not as interesting (e.g., color or smell). Since they are so important, we pay special attention to the mechanical properties of materials.

Mechanical properties are design criteria that you would find in a set of specifications—they deal with the load and deformation limits of a material, such as strength, stiffness, and extensibility. Mechanical properties describe how a material should perform in the field and are key parameters for structural engineering. **Chapter 2** focuses solely on the mechanical properties of materials. To establish these properties, we measure a material's performance in the lab under a series of tests that might push, pull, twist, deform, age, or generally damage the material in an attempt to find its limits. Examples of mechanical properties include strength, formability, stiffness, toughness, and durability (see **table 1.1**).

Nonmechanical properties can be broken into four subgroups: physical properties, chemical properties, transport properties, and process properties.

a. **Physical properties** describe how a material looks or feels with respect to the laws of physics. They include properties such as shape, molecular structure, color, opacity, roughness, and mass. These are properties that can usually be measured through nondestructive testing and/or microscopic observation.

b. **Chemical properties** are characteristics relating to a material's atomic makeup and reaction potential; they define how a material will chemically react in the design environment. It is important to make sure that a chosen material is environmentally safe, meaning that it does not degrade or corrode because of the environment or pollute a surrounding environment.

TABLE 1.1 PROPERTY EXAMPLES

Property Type	Property Subgroup	Example(s)
Mechanical	N/A	Strength
		Formability
		Stiffness
		Toughness
		Durability
		Creep
		Ductility
		Modulus of elasticity
		Elongation
		Proportional limit
Nonmechanical	Physical	Shape/geometry
		Texture
		Aesthetics
		Unit weight density
		Specific gravity
		Permeability
		Moisture content (MC)
		Porosity
	Chemical	Corrosion resistance
		Electrochemical reactivity
		Ultraviolet (UV) susceptibility
		Flammability
		Biodegradation
	Transport	Electrical conductivity
		Thermal conductivity
		Ion exchange
		Gas diffusion
	Process	Resonance
		Emissions
		Acoustical masking

c. **Transport properties** are related to processes that move a quantity through a material, such as electric current or heat, and depend on "moving ions." They are related to chemical properties but can be separated due to their importance.

d. **Process properties** describe how materials or material systems physically interact with their surroundings. Vibration dynamics is an example of a process property, which is important to understanding how a material will respond in an earthquake or dampen acoustical noise in a concert hall.

Currently, the most common materials used in construction by civil engineers are steel (**chapter 5**), aggregates (**chapter 6**), concrete (**chapter 6**), asphalt (**chapter 6**), wood (**chapter 7**), soil (**not covered**), and masonry (**not covered**). While many of these materials have been around for a while (the Bedouins first used concrete in 6500 BC!) and will probably continue to be used into the future, we have seen and will see vast improvements in their performance. Additives such as fiber reinforcement and super plasticizers in concrete are just two examples of chemical and nanoscale research improving material performance.

In recent years, there has also been significant growth in the use of **high-performance materials**. Advanced polymer composites, geotextiles, lightweight metals (like aluminum or titanium), and engineered woods (glulam) are just some of examples of new materials being selected by engineers. Joining the high-performance materials revolution is also another movement—the call for **sustainable or "green" materials** (revisited in **chapter 8**). In an effort to create sustainable designs, engineers are selecting materials that will reduce energy consumption and potentially improve the environment. To achieve this, engineers could select recycled or recyclable material to minimize the energy required to manufacture a new material or select a great insulating material to minimize energy transfer from the building to the environment.

IV. Illustrating Physical Properties: Mass, Density, and Weight

Civil engineers like to describe a material with respect to its physical structure and the nature of its matter. Of particular importance is the degree of packing of the material; there are three general ways of measuring this. The first, **density** (ρ), is defined as the mass per unit volume and is given by the following equation:

$$\rho = \frac{\text{mass}}{\text{volume}} = \frac{m}{V}.$$

The second, **unit weight** (γ), is defined as the weight of a material per unit volume, or

$$\gamma = \frac{\text{weight}}{\text{volume}} = \frac{W}{V}.$$

The equations for density and unit weight are related to each other through **gravity** (**g**). The major determining factor when using either measurement is the system of units with which you are working (metric or imperial). The following equation shows the relationship:

$$\text{weight} = \text{mass} \times \text{gravity},$$

or simply,

$$W = mg.$$

From the definition of unit weight, we can easily derive a relation between unit weight and density, where

$$\gamma = \frac{W}{V} = \frac{mg}{V} = \left(\frac{m}{V}\right)g = \rho g.$$

Thus,

$$\gamma = \rho g.$$

The third most common way of describing a material is to designate its specific gravity. **Specific gravity** (**G**) is defined as the ratio of a material's mass for a given volume to an equivalent volume of water, both at the same given temperature, T°, or

$$G = \frac{\text{density of material @ } T^\circ}{\text{density of water @ } T^\circ} = \frac{\rho_{material}}{\rho_{water}}.$$

Since specific gravity is a ratio measurement, we could interchangeably say that

$$G = \frac{\text{unit weight of material @ } T^\circ}{\text{unit weight of water @ } T^\circ} = \frac{\gamma_{material}}{\gamma_{water}},$$

as

$$G = \frac{\rho_{material}}{\rho_{water}} = \frac{\rho_{material}}{\rho_{water}}\left(\frac{g}{g}\right) = \frac{\gamma_{material}}{\gamma_{water}}.$$

As a result, G is unitless and is constant regardless of using metric or imperial units.

With such relations, we can begin to define material properties and objectively compare them. For example, aluminum has a specific gravity of 2.7, while steel has a specific gravity of 7.8. *Does that mean that those materials will sink or float in water? What is the specific gravity of water?*

Working out the mass and/or weight of a material is relatively easy (only requiring a set of scales). However, such properties related to mass do not tell us if the material is suitable for a *particular function*. Is a dense material strong or weak? We don't know! To answer such questions, we need to assess properties, such as **strength** and **stiffness**. How does one go about measuring strength? Today, with various testing equipment and machines, such tests seem obvious. But this was not always the case. The logical step was applying the **scientific method** to materials testing, which we will briefly discuss to help understand how material properties are determined.

V. The Scientific Method

The so-called scientific method is a formal description of scientific inquiry that has characterized natural science since the seventeenth century. In effect, if you have a question that necessitates a scientific answer or explanation, certain steps *must* be followed to be accepted by the scientific community. There are many ways to describe the scientific method (which we discuss in more detail next), but the key components are (1) a well-defined question or **hypothesis**, (2) an **experiment** to test the hypothesis, and (3) results to support the hypothesis. This may seem obvious, but it was not always the *status quo*. To illustrate, we can consider the famous case of Galileo's so-called Leaning Tower of Pisa experiment:

In 1589, the Italian scientist Galileo Galilei is said to have dropped two spheres of different masses from the Leaning Tower of Pisa to demonstrate that their time of descent was *independent* of their mass. Via this method, he supposedly discovered that the objects fell with the same acceleration (Earth's acceleration owing to gravity, g), proving his prediction (or **hypothesis**) true, while at the same time disproving the long-believed theory of gravity proposed by Aristotle (which states that objects fall at a speed relative to their mass—i.e., heavier objects were thought to fall faster). Now, the idea that heavier objects fall faster may make sense if you compare something like a balloon and a brick.[3] However, no one *actually tested* this fact to make sure it was universally true. Galileo did with a simple experiment.[4]

If you remember equations of motion from physics, you know that an object that starts from rest with a constant **acceleration**, a, will cover a **distance**, x, over a **time**, t:

$$t = \sqrt{\frac{2x}{a}}.$$

---

3   The balloon may fall slower because of the air resistance or *drag*, but that has nothing to do with its *mass*.

4   While this story is commonly retold as historical fact, there is no account by Galileo himself of such an experiment, and it is accepted by most historians that it was only a *thought experiment* that did not actually take place—a *Gedankenerfahrung*.

It is clear from the math that *mass does not influence the time*! These equations are supported by the results of Galileo's experiment—we trust them because of the scientific method. They are physically testable!

The scientific method provides information to support a claim or hypothesis. From a philosophical sense, there are no definitive answers. This is, however, an important concept in materials science, because it effectively means all our *testing provides approximate answers.*

Wait, what does this mean?

If we test steel and get a strength of 30,000 psi, we typically report that "steel has a strength of 30,000 psi." However, what is unwritten is that "the particular steel specimens we have tested have an average strength of 30,000 psi within a range of precision limited by our testing equipment." The more *support* for the number (30,000 psi), the more confidence you can have in the property. More support comes from repeated testing (ideally using different methods and from independent groups). However, you can never know 100 percent that the *next* steel specimen will have the same strength—which is why engineers always apply conservative safety factors to material properties.

The scientific method is commonly based on empirical or measurable evidence subject to specific principles of reasoning, typically accomplished in a stepwise manner. Laboratory exercises typically follow such a procedure. However, the steps are frequently modified to suit the needs of the problem, so there is no "definitive guide to the scientific method." With that being said, one *interpretation* of the steps in the scientific method consists of the following:

1. **Ask a question:** What is the basic problem you are solving? Or, from a material perspective, what property, characteristic, or behavior are you investigating?
2. **Do background research:** Has anyone tackled this problem before? Or something similar? What is already known?
3. **Formulate a hypothesis:** How do you think the experiment will result and why? Make a prediction. Don't worry if you are right or wrong; that's what the test is for!
4. **Design an experiment:** Develop a procedure to test your hypothesis or determine a material property (see the discussion on experimental design in the next section).
5. **Perform the experiment and collect data:** Gather and record all relevant data, measurements, and observations.
6. **Interpret data:** Analyze data to aid in theoretical descriptions.
7. **Draw conclusions:** Was your hypothesis correct? Did you determine a material property or behavior? Are the conclusions adequately supported independently by the results?

8. **Refine the procedure** (optional): If the hypothesis failed or the experiment was insufficient, revise and repeat steps three to seven.
9. **Communicate the results:** Summarize findings to add to "science."

The final step (**communication**) is critical, because a single test/result does not prove anything. Results are validated and vindicated when other scientists repeat experiments and come up with the same (or supporting) results. A history of evidence and validations shows that the original statements were correct and accurate. It is a simple idea and the basis of all science—*validation through replication*. Statements must be confirmed with as much evidence as possible. In fact, failure to repeat results has led to a so-called replication crisis in academia. This **replication crisis** refers to alarming (and growing) evidence scientists have found that indicates that the results of many scientific experiments are difficult or impossible to replicate on subsequent investigation, either by independent researchers or even by the original researchers themselves. Since the **reproducibility** of experiments is an essential part of the scientific method, this phenomenon potentially has grave consequences for many fields of science in which significant theories are grounded on experimental work that has now been found to be resistant to replication. Luckily, the crisis is more prevalent in the fields of biology and medicine, not materials science or engineering.[5]

Perhaps the most critical step in the scientific method (and for replication) is prudent and careful consideration of experimental design. How you *attain* data is almost as important as the data itself! As such, **experimental design** is the next topic of discussion.

VI. Experimental Design

Experimental design consists of the design of a procedure or task to describe or explain some variation of information/data. Planning an experiment properly is very important in order to ensure that the *right type of data* and a *sufficient sample size* are available to answer the research questions of interest as clearly and efficiently as possible. There are many interpretations of experimental designs, but for our purposes, we consider material testing a primary goal.

From a material characterization perspective, experimental design effectively means the design of a test procedure to determine a material property or behavior. Double-blind placebo studies are not what we are interested in. If we want to know the strength of steel, for example, we need to test it until it breaks. Seems simple enough at first glance.

More formally, the key concepts of experimental investigations are (in the context of steel testing) (a) **manipulation** of a desired independent variable (e.g., increase load on steel

---

5   It turns out that it is relatively easy to break things consistently!

specimen); (b) **control** of other variables, except the dependent variable constant (e.g., constant specimen geometry); and (c) **observation** of the independent variable on other dependent variables (e.g., measure change in length with respect to the change of load). Specific questions that the experiment is intended to answer must be clearly identified before carrying out the experiment. If our goal is testing the strength of steel (defining the output), for effective experimental design, we need to ask a set of simple questions:

1. What will our test specimens look like in terms of geometry?
2. How many specimens will be needed?
3. Will they all be tested to failure?
4. How is failure defined or observed?
5. What equipment will we be using?
6. What kind of load will be applied (stretching, squeezing, twisting)?
7. How will we gather the data? What data?

Those initial questions will be a good start and may result in a simple test for the strength of steel (sticking to our initial example). In effect, we are defining the **controlled experimental variables**.

Things will get complicated as you delve a little deeper into the test details and other questions emerge. Such questions include the following: Do I have enough specimens to be statistically relevant? Do I need control specimens? Does geometry matter? Does loading rate matter (e.g., how fast you test a specimen)? Does the temperature matter? Humidity? Frequency of data collection? Precision of the measurement? Unfortunately, some of these questions *cannot* be answered until you perform some **trial experiments**. One should also attempt to identify known or expected sources of **variability** in the experimental units since one of the main aims of a designed experiment is to reduce the effect of these sources of **uncertainty**.

A well-designed experiment is logical, delivers the predicted data and/or results, and can be easily replicated if the procedure is followed under the same conditions at a different time.

In theory, one can design multiple tests and attempt to characterize many different materials. However, if we design a specific test, and you design a specific test, will we get the same results? Unless we collaborated with each other, it is likely (a) our tests may be completely different, requiring different equipment, specimens, and so on, and (b) the results may slightly differ because of slight variations in the test procedure. If they differ, whose results should be trusted? Whose experimental procedure was "better"?

This becomes an issue for commonly used (and thus frequently tested) materials. Common ground is needed between testing parties. Rather than "reinventing the wheel" every time a material property needed to be assessed, a more consistent method is required.

Luckily, many engineers in the past have determined what properties are necessary for particular functions and developed the "best" ways to test for these properties.[6] These methods became so-called standards (i.e., a set of guidelines judged to be suitable to assess materials). To make things more consistent for design and to ensure public safety, a formal system of **material standards** enables property selection based on known values grounded on prior accepted tests. Such standards also dictate extremely specific procedures for measuring the properties of materials, thereby allowing comparison of results from any source as long as the specific procedure is followed.

VII. Material Standards

Material standards are generally accepted and theoretically sound rules to determine the quality, performance, and proper use of a material. Material standards are usually determined by external agencies instead of individual manufacturers. Every standard usually focuses on a specific, measurable concept and is clearly written to ensure consistency and to avoid conflict within a family of standards. Well-written standards (usually those with the broadest marketplace acceptance) help industry by improving products, expanding markets, and lowering risk. Standards can be used by individuals, companies, and even government agencies, which all reference them for guidance. In fact, enactment of the **1996 National Technology Transfer and Advancement Act** (Public Law 104-113) required government agencies to use privately developed standards when at all possible! This law has proven beneficial for two reasons: it has ensured the use of high-quality, market-relevant standards and has also saved taxpayers millions of dollars by eliminating duplicative standards development efforts.

In the United States, there are over 400 industry-driven standards development organizations, while most other countries use a government-based approach. This is not to say that there are not government-based standards agencies in the United States. Examples of government standards range from local (e.g., town building codes) all the way to the federal level (e.g., **National Institute of Standards and Technology** (**NIST**), to name one). The largest private standards organization in the United States is **ASTM International** (formerly known as the American Society for Testing and Materials). There are five main standard types recognized by ASTM International, including the following:

1. **Product specification standards**—which govern how things are called for in a procurement document

---

6   What makes them the "best"? Think about it.

2. **Manufacturing practice standards**—which govern protocols that do not produce a test result
3. **Test method standards**—which govern lab and field tests that produce a specific result
4. **Standardized terminology standards**—which set official definitions
5. **Operational and purchasing guide standards**—which give informational descriptions of options

To ensure that buyers know that they are purchasing an item that has met a particular standard, many items are stamped with their relevant standard number. **Figure 1.1** is an example of a pipe in the Hurtig Building at Northeastern University (Boston, Massachusetts) that displays an ASTM stamp. Manufacturers of products following any of the internationally recognized standards are more likely to be used, as they are more likely to be called for in a contract. This is in part because of a large movement by international trade organizations, such as the World Trade Organization, to encourage the use of standards to support fair trade practices. (Fair trade is defined herein as a trading condition that is sustainable both environmentally and economically.) Standards help ensure that when a material is specified in a contract, all parties involved benefit from uniformity and transparency.

As another example, **figure 1.2** shows a standard sieve, which is used for the Standard Test Method for Sieve Analysis of Fine and Coarse Aggregates (ASTM C136/C136M-14). The test is used to determine the particle size distribution of a particular aggregate of interest.

You may think that all sieves are alike—just grab one from a sandbox or a baker's kitchen! This, however, is a recipe for disaster (or at least inadmissible testing)! Precision sieves

FIGURE 1.1   Example of ASTM labeled pipe.   FIGURE 1.2   Sieve.

with standardized materials are an essential part of the sieve analysis, and an incorrectly fabricated mesh or ill-advised material can result in bad and/or useless test results. ASTM designation E11-17 states,

> The sieve cloth used in test sieves shall meet the requirements of Table 1 and shall be designated Specification E11 Sieve Cloth. The number of inspected apertures shall be in accordance with Table 1 (Column 7). Sieve cloth conforming to this specification shall be woven from stainless steel, brass, or bronze. Sieve cloth with openings greater than or equal to 75 micrometers shall be woven using a plain weave. For sieve cloth with openings equal to or less than 71 micrometers the sieve cloth may be supplied using a twill weave. The sieve cloth shall not be coated or plated.[7]

When a difference exists between the tool and the test specification, there is a lot of space for variance in your results. Simply put, if you do not use the "standard" sieve, you did not complete the "standard" test, and you may have just as well not tested at all.

One *negative* aspect of standards is that *meeting them* is frequently a significant expense. For example, the sieve depicted in **figure 1.2**, while appearing equivalent to a baking gadget or kid's toy, cost more than $50 to purchase. While meeting standards adds value to a material or product, it is not free to subject materials to testing. Moreover, the testing equipment itself is often highly specialized (as it must meet the minimum requisite measurement precision). Likely, materials and products are sent to third parties who specialize in testing procedures. This ultimately increases the cost of the material. Guaranteed quality and performance costs money. It is one (of many) **economic factors** that must be considered in **material selection** for engineering design and application.

VIII. Economic Factors

There are many factors involved in the **economics of materials selection**.[8] Not only must we consider how much a material costs at the time of construction, but also what the costs of the material will be over its *lifetime*. Life-cycle costs

The following link will take you to the ASTM page for the standard sieve test. Please note that you must have a university library with a subscription in order to access.

https://compass.astm.org/EDIT/html_annot.cgi?C136+14#s00013

As you can see, the details are quite specific for every aspect of this test right down to the test scale tolerances and where the mesh midpoints are measured! This ensures true uniformity for anyone performing the test in the United States or abroad and results in product reliability.

---

7   ASTM, *Standard Test Method for Sieve Analysis of Fine and Coarse Aggregates, C136/C136M-14* (West Conshohocken, PA: Author, 2014).

8   A fancy expression for the cost of materials. Money is the bottom line!

are becoming increasingly important to owners and affect the selection of materials. This is due to smarter financial planning by many owners and can also be accredited to growing environmental and social movements.

The process of accounting for costs over a material's lifetime is known as **life-cycle cost analysis (LCCA)**. LCCA includes the following: (a) raw material acquisition and material manufacturing processes, (b) transportation expenditures from every phase of the material's life, (c) the installation and use in a structure (including ongoing maintenance), and (d) end-of-life disposal via recycling or being placed in a landfill.

There are many "hidden costs" associated with every material. Here, our discussion just scratches the surface with the most apparent costs in a material's "life." The **environmental cost** of extracting many raw materials is very high, with pollution a major weighing factor. Moreover, **raw materials** from oil to iron to wood can vary in availability because of political issues or global climate change. **Figure 1.3** depicts a graph with the pricing of crude oil (which relates to **transportation costs**), with some major international events highlighted. The international conflicts affected the availability and cost of crude oil, which in turn increased the costs of many raw materials.

Once a material is selected (made either from raw or recycled components), **manufacturing costs** must also be considered. The typical construction site does not possess the ability to turn raw limestone and shale into future concrete and must rely on manufacturing plants to provide a finished product at reasonable prices. Should **market demand** for the

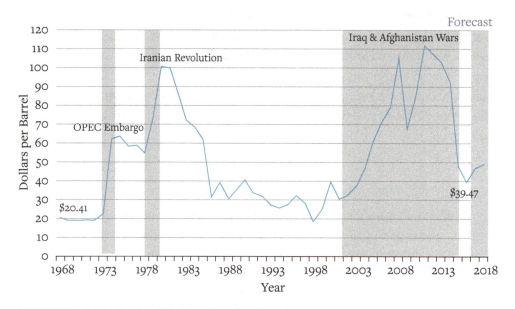

FIGURE 1.3   Crude oil prices (U.S. dollars/barrel), 1968–2016.

finished materials be high, manufacturers can take advantage of supply-and-demand economics and adjust their pricing. It should come as no surprise that manufacturing costs for many construction materials may fluctuate into the future as developing nations such as China, India, and various African countries begin developing their infrastructure systems (see **figure 1.4**). It is at this intersection of interests that new construction products may find themselves at the forefront of industries struggling to keep up with demand.

In selecting materials, you must also consider the **transportation costs** of getting the material to your site. Whether it is transportation from a different coast or a different continent, your material needs to arrive at your site on time. Timely delivery requires knowledge of the manufacturing and shipping schedules of a material—called the **lead time**. The cost to

FIGURE 1.4   Construction of a new highway in Accra, Ghana, in 2010. Photo credit: Adrian Brügger.

transport materials expediently may make a cheaper material less economical when rush transportation costs are considered. **Rush costs** are higher not only because they take advantage of need, but also because they cannot take advantage of economies of scale; a single item must bear the totality of the transportation cost. Also included in transportation costs are **road access costs**. Large and/or heavy items, such as precast concrete and irregularly long steel beams, may require special oversize-load permitting and police details to deliver them to your site by truck.

The next costs to consider in material selection are **placing costs. Skilled labor** must be available to place the material. For example, specially trained laborers are required for working with steel and constructing tall structures in large cities. If a plan called for a large steel structure in a smaller town, not only could the number of laborers available to work be inadequate, but also those available may not have experience working on tall steel structures. The solution would be to hire skilled labor from a nearby region and put them up in hotels—thus significantly raising your placing costs. Other placing costs might include seasonal weather considerations. For example, significant effort must go into ensuring a proper cure when placing concrete in a New England winter. Heaters and blankets may be employed, leading to raised placing costs.

Once your structure is in place, your client may want to know about the remaining life-cycle costs, including the maintenance and end-of-life costs. **Maintenance costs** for materials might include patching for concrete, structural weld inspections for steel, and termite inspections for wood; maintenance costs may also take into account time to replace a given material. **End-of-life costs** look at the cost to remove a material from a site. These costs can range from landfill costs to recycling costs, depending on the material selection. Many designs currently lean toward the selection of materials that can be used in endless cycles of building, thus lending themselves to construction that touts environmental responsibility in material selection and sustainability; what does sustainability entail?

IX. Materials and Sustainability

A good synopsis of sustainability is "don't do today what will damage tomorrow." In this text, **sustainability** will be framed using the American Society of Civil Engineers (ASCE) designation:

> ASCE defines sustainability as a set of environmental, economic, and social conditions—the "Triple Bottom Line"—in which all of society has the capacity and opportunity to maintain and improve its quality of life indefinitely, without degrading the quantity, quality or the availability of natural, economic, and social resources.[9]

Sustainability and related concepts and practices make sense from current ethical perspectives and social pressures. Surprisingly, this was not always the case.

Historical Perspective

For centuries, sustainable elements have been included in construction (usually by serendipitous coincidence and lack of options, rather than any intent or sense of responsibility). Historical examples of sustainable elements include the use of local and renewable materials, passive solar heating, and the use of landscaping for temperature control. It is only in the last 200-plus years that design and material selection have changed so rapidly because of accelerated material design and discovery, alongside manufacturing innovations. With global trade, we have been able to obtain inexpensive materials from across the world. However, as materials are brought from all the corners of the globe for economic benefit, they bring with them environmental detriment in the form of pollutants created by their extraction, transportation, and processing. One common measure of this environmental detriment is an object's **carbon footprint**.

A material's carbon footprint is best described as the **greenhouse gases** (GHG) expended to acquire, process, transport, and finish a particular material. One example of a finished material with a large carbon footprint (and thus not a very sustainable one) could be a road

---

9 American Society of Civil Engineers (ASCE), "Sustainability," https://www.asce.org/sustainability/.

in Kansas made with bitumen from Venezuela and aggregate from Canada. While the initial costs may have been low for this hypothetical road, the environmental and whole life-cycle costs may tell a different story.

While local, minimally processed materials may have been the only options in the past, the age of fossil fuels has enabled the selection of cheaper and better-performing materials. However, cost and performance are only part of the picture, which needs to include environmental factors in order to create a full picture of a material's life cycle.

Contemporary Times: Energy Efficiency and Environmental Protection

The current **environmental movement** was born out of the industrial or manufacturing revolution of the eighteenth and nineteenth centuries, for better or for worse. During the **Industrial Revolution**, technology

MODEL OF A GREAT SUSTAINABLE BUILDING

The Colosseum in Rome is a great example of a building with sustainable materials. The main structural pillars and external walls (still standing today, as shown in **figure 1.5**) were made from limestone from Tivoli, only 20 kilometers away. At the end of the structure's useful life, stone from the façade was recycled to make the steps of the city's churches, including those of St. Peter's Church, and 300 tons of metal clamps (used to hold the decorative stones) were recycled to make weapons in the Middle Ages. The evidence of this building's continued existence almost 2,000 years later, and its ability to be reused over the years, certainly speaks to the sustainability of its construction!

FIGURE 1.5   Colosseum in Rome.

improved rapidly because of the use of fossil fuels, such as coal and oil. Many businesses set up shop in densely populated cities and traded goods globally as shipping became faster with fossil-fuel-burning technologies. Concurrently, standards of living improved, but human health issues related to a declining ecology also increased. More air and water pollution from fossil-fuel-burning activities led to private environmental protection societies and public land preservation acts in the early 1900s. By the 1960s, the environmental movement was a national issue in America, leading to the enactment of federal legislation protecting public health and land, and from which we still benefit today. Noteworthy legal actions included the Clean Air Act of 1963, the founding of the United States Environmental Protection Agency (EPA) in 1970, the Endangered Species Act of 1973, the Safe Drinking Water Act of 1974, and the Comprehensive Environmental Response, Compensation and Liability Act (CERCLA) of 1980 (also known as the Superfund Act). These new laws reflected the reality that industrial and technological "progress" came at an environmental price and that our air, water and land was in need of protection.

While human health concerns led to the environmental movement, among the most powerful catalysts in bringing sustainable energy practices to the public's attention were the Arab oil embargoes of 1973 and 1979. During this time, the price of oil skyrocketed, affecting nearly all aspects of the global economy. The tangible effect on people's wallets finally motivated political action. In response to the economic upheaval created by the embargoes, an **energy efficiency movement** arose. Several energy efficiency measures were passed by Congress, including a change in fuel economy standards from 13.5 miles per gallon (mpg) to 27 mpg and the enactment of the maximum speed limit law, which forced vehicles to travel at a maximally efficient speed.

Taking cues from the environmental and energy efficiency movements, the contemporary **sustainable building movement** arose from the need for more energy-efficient and environmentally friendly building practices. These practices called for design mindful of energy consumption and environmental impact for both new construction and the retrofitting of existing structures. According to the United States Environmental Protection Agency (USEPA), sustainable building is

> the practice of creating structures and using processes that are environmentally responsible and resource-efficient throughout a building's life-cycle from siting to design, construction, operation, maintenance, renovation and deconstruction. This practice expands and complements the classical building design concerns of economy, utility, durability, and comfort.[10]

In effect, such concerns placed civil engineering at the forefront of sustainable practices.

Driving Agents for Sustainability in Civil Engineering

As civil engineers, we are involved in the sustainable building movement through design, construction practices, and selection of sustainable materials. For example, a sustainable building could

- be designed to incorporate materials that create healthy indoor environments with minimal pollutants (e.g., reduced product emissions);
- be constructed with a waste management plan that minimizes the waste stream and maximizes material reuse; and

---

10   United States Environmental Protection Agency (USEPA), "Green Building," February 20, 2016. https://archive.epa.gov/greenbuilding/web/html/about.html.

- employ structural and finishing materials that can be reused, contain recycled content, or are made from renewable resources.

In 1996, the ASCE code of ethics was modified to include sustainable development as part of the canon of civil engineering principles. As the leading society for the profession, ASCE has brought recognition and significance to the sustainable building movement. In changing the ASCE canon, the organization demonstrated its conviction that the principle of sustainable development was of such importance as to be before all others, as shown in **table 1.2.**

TABLE 1.2 **ASCE FUNDAMENTAL CANONS**[11]

1. Engineers shall hold paramount the safety, health, and welfare of the public and shall strive to comply with the principles of **sustainable development** in the performance of their professional duties.
2. Engineers shall perform services only in areas of their competence.
3. Engineers shall issue public statements only in an objective and truthful manner.
4. Engineers shall act in professional matters for each employer or client as faithful agents or trustees and shall avoid conflicts of interest.
5. Engineers shall build their professional reputation on the merit of their services and shall not compete unfairly with others.
6. Engineers shall act in such a manner as to uphold and enhance the honor, integrity, and dignity of the engineering profession and shall act with zero tolerance for bribery, fraud, and corruption.
7. Engineers shall continue their professional development throughout their careers and shall provide opportunities for the professional development of those engineers under their supervision.
8. Engineers shall, in all matters related to their profession, treat all persons fairly and encourage equitable participation without regard to gender or gender identity, race, national origin, ethnicity, religion, age, sexual orientation, disability, political affiliation, or family, marital, or economic status.

One of the principal contemporary drivers of the sustainable building movement is the **US Green Building Council** (USGBC). The USGBC is best known for its **Leadership in Energy and Environmental Design** (LEED) program, formed in 1998. This program employs third-party verification (clients must hire LEED-certified engineers or architects to prepare documentation for submission to USGBC) to certify buildings, homes, neighborhoods, and communities as sustainable. USGBC has various certification levels, depending on the degree of sustainability in a design. Some major areas that would qualify a building as sustainable are described in **table 1.3.**

---

11   ASCE, "Code of Ethics."

TABLE 1.3 SUSTAINABLE BUILDING CONSIDERATIONS

Area	Attained by
Reduction of GHG emissions	Selection of local and/or low CO_2 materials (reduce embodied GHG emissions)
	Energy efficient design (reduce operational GHG emissions)
Waste-stream reduction	Construction waste-management plans
	Construction sedimentation reduction plans
Water-waste reduction	Water-use reduction inside the building
	Efficient landscape design aimed at reducing water need
	Minimization of rainwater runoff
Health and safety for occupants	Materials that reduce toxin exposure to end users
	Climate controls, ventilation, optimization of daylight exposure
Site selection	Choosing a site in a well-connected area for public transit and biking
	Designing a site to minimize environmental impact

How Significant Is a Material's Life Cycle?

The most important theme in designing sustainable buildings lies in addressing their entire life cycle. As we discussed earlier in the chapter, LCCA looks at all of the costs that are associated with an item, from beginning to end. Similarly, we can also systematically account for environmental impacts through the life cycle of a material through a **life-cycle assessment** (**LCA**) (further discussion in **chapter 8**). Like LCCA, a material's LCA examines environmental impacts accompanying all phases of a material's life, from cradle to grave. These phases include (a) raw material extraction; (b) transportation over all phases of the material's life; (c) the energy, chemicals, and impacts associated with manufacturing processes; (d) the installation and use of the material in a structure; and (e) material disposal via recycling or placement in a landfill.

For a building, the life cycle encompasses the idea of the building itself, as well as all of its individual material components. In conducting the LCA of a material, we must first consider how the raw material is acquired. We will consider the extraction and production of aluminum as our example. It should be noted that aluminum has relatively high chemical reactivity, which means it tends to bond with other elements to form compounds and thus is found within other elements.

The point of extraction can affect surrounding ecosystems through dispersion of emissions and wastes into air, land, and water. This phase involves many fossil-fuel-burning machines to extract the raw material, bauxite, from large open pits, as shown in **figure 1.6.**

FIGURE 1.6   Example of an open pit mine.

Once the raw material is extracted from a location, it is transported to a primary processing plant for processing and refining. This stage is the most highly toxic phase and involves massive chemical and energy inputs, as illustrated next.

Raw aluminum + 12 kg of input materials + 290 MJ = > 15 kg CO_2 emissions + 1 kg Al

Inspecting the earlier equation, aside from the fact that 1 kilogram of aluminum (Al) results in 15 kilograms of carbon dioxide (CO_2), some questions to consider include the following: What happens to the 12 kilograms of input materials? Can it be reused? Is 290 MJ a lot of energy? What is the energy source? How is it generated? Did that produce more CO_2? More waste? Down the rabbit hole we go. Typically, the more questions asked, the worse a material becomes.

Lastly, the material is transported to another facility for manufacturing and assembly into products, such as aluminum beams. This stage of the process might involve high-temperature ovens to melt the aluminum into the desired shapes and can involve the use of oils, solvents, coatings, and other alloying metals. Surprisingly, this phase is the least

TABLE 1.4   ENVIRONMENTAL CONCERNS CONNECTED TO MATERIAL SELECTION

Concern	Materials Link
Climate change	• Emissions from materials manufacturing • Transportation of materials • Landfill gas emissions from disposal of material
Fossil-fuel depletion	• Creation of plastics, asphalt cement, solvents, sealants, direct/indirect use of fuels to manufacture, process, and deliver materials
Air pollution	• Fossil-fuel combustion during materials' lifetime • Emissions from material cleaning and cooling processes
Deforestation, desertification, and soil erosion	• Commercial forestry and agriculture • Resource extraction • Mining • Dredging
Water pollution and eutrophication	• Mining nutrient drainage • Water use and effluent discharge from material manufacturing process • Construction site fine sediment runoff • Waste disposal
Energy waste	• Excessive power and heating needed because of poor thermal conductivity of a selected material

polluting when compared to the primary processing. While the material is now ready for transportation to a construction site for use, our LCA does not end here. Most buildings are generally designed for 50 years of use. The final point to consider in the LCA is whether the aluminum can be easily reused or recycled at the end of its life.

Through this example, we see the environmental costs are great when we are considering using new aluminum as material. From extraction to landfill, there are many chemical and energy inputs needed to make this material. Recognizing the inputs in this complete life cycle allows us to understand the need for sustainable building practices that call for recycled materials. By employing reused and recycled materials, and eliminating the primary processing step, we can also eliminate a significant portion of the associated environmental maladies. Consider the environmental impacts and effects of materials processing outlined in **table 1.4**.

X. Concluding Remarks

In conclusion, this introductory chapter described the importance of materials in engineering as well as the requisite material knowledge to use them reliably in the built environment.

To understand materials fully, we need to determine their properties, such as density and specific gravity, which requires scientific inquiry and associated experimental design. Such explorations define the process known as the scientific method. Need for consistency led to the development of material standards, which aids in the selection of materials and regularity from application to application. Materials selection is the process of weighing many factors, from the performance desired and the restrictions of budgeting and economic factors to selecting an environmentally responsible material and weighing ecological affects (i.e., sustainability). A successful engineer must take into account all considerations to make the most apt choice of materials for design. Now, we proceed to define some key material properties and behaviors—specifically, how to describe the underlying **mechanical principles** and responses of common engineering materials.

XI. Problems

1. What are the characteristics of good engineering materials?

2. Identify a material that is NOT a good engineering material and explain why it is not a good engineering material.

3. In the context of materials, what does the term "workable" imply?

4. Describe the difference between mechanical and non-mechanical properties.

5. Match each material property to the category that best describes it.

Categories	Material Properties:
a. Mechanical	Resonance
b. Physical	Flammability
c. Chemical	Thermal conductivity
d. Transport	Corrosion resistance
e. Process	Creep
	Specific gravity
	Strength
	Ultraviolet susceptibility
	Electrical conductivity
	Shape

6. It is proposed to use piles to support docks in saltwater at a marine port facility. Name one property in each of the following categories that is important for

the pile material in this application. Identify a material that may work well in this case.

    a. Physical property

    b. Mechanical property

    c. Chemical property

7. A metal rod is 2 cm in diameter and 10 cm long. In the lab, the mass of the rod was measured at 84.76 g. The temperature of the rod in the lab is 20°C. Using this information, determine the following:

    a. The density of the metal.

    b. The unit weight of the metal.

    c. The specific gravity of the metal (the density of water at 20°C is 0.998 g/cm^3).

    d. Based on the prior results, what metal is the rod most likely made from?

8. A block of wood has a volume of 150 cm^3. Its mass was measured on a balance in the lab at 135g.

    a. What is the weight of the block of wood, in Newtons?

    b. What is the density of the wood?

    c. What is the unit weight of the wood?

    d. What is the specific gravity of the wood (you may assume the density of water is 1g/cm^3)?

9. In a few sentences, explain what a standard is and why standards are important in engineering.

10. Look up the following ASTM standards. Give the title of each and summarize them in one or two sentences each.

    a. D2525

    b. C134

11. Write a pseudo standard for exterior deck screws that includes at least three criteria.

12. Identify at least four different costs associated with materials that impact material selection and design.

13. Find a sustainability rating system (other than the US Green Building Council and the LEED program for certifying sustainable buildings) and report the following:

    a. Who developed the system.

    b. The applicability of the system (e.g., LEED is primarily for buildings)

    c. How certification in the system is administered/evaluated

14. Many materials are manufactured from raw materials that are obtained through mining. Identify at least two environmental concerns that are directly associated with mining.

2

Mechanical Principles

I. Introduction—Why Mechanics Is Important for Materials

One take away from **chapter 1, "Introduction to Engineering Materials,"** is simply that civil engineers build things from materials, and in order to do so, we need to know the material properties and behaviors. Engineers must be able to predict how the final built system will behave when subjected to loads. For example, how much will a bridge sway under high winds? Relating loads (or forces) and the resulting movement (or displacement) is one of the fundamental requirements of civil engineering design, critical to skyscrapers, bridges, highways, or foundations equally. This involves a keen understanding of the field of **mechanics**.

Mechanics is broadly defined as the branch of science concerned with the behavior of physical bodies when subjected to **forces** (typically denoted with an F) or **displacements** (denoted with the Greek letter delta, or δ) and the subsequent effects of the bodies on their environment. For example, applying **equilibrium** conditions (from any introductory **statics** class), we find the following:

$$\sum F = 0.$$

For example, the sum of all forces (or *net force*) acting on an object equals zero (*if* the object is *static* or *not* subject to acceleration). The earlier equation is a simple application of **mechanics**—the loads in a system must be in equilibrium if the system is to remain at rest. This is a simple variation of **Newton's first law of motion**, which states that an object will remain at rest (or in uniform motion in a straight line—e.g., *constant velocity*) unless acted on by an external force.

If the forces are not balanced (i.e., equilibrium is not attained), we must consider **Newton's second law of motion,**[1] or

$$F = ma.$$

---

1   Newton's second law of motion pertains to the behavior of objects for which all existing forces are *not* balanced. The second law states that the *acceleration* of an object is dependent on two variables: the *net force* acting on the object and the *mass* of the object.

In this case, the unbalanced force (F) causes a mass (m) to accelerate (a) in the direction of the force—it *moves*! Most structures in civil engineering are designed not to move (we can joke that "if it moves, it's broken"), so we are primarily concerned with **equilibrium conditions**[2] (however, accelerations do come into play and must be considered in scenarios such as earthquake design).

We can also combine the principle of equilibrium with **Newton's third law**, which states that for every action there is an equal and opposite reaction. Thus, setting a net force to zero for a given system, and then applying the principle of action/reaction, enables us to solve for reaction forces in a simple structure, such as a truss. Note that these laws hold *regardless of the material* and that Newton's work formed the basis of classical engineering mechanics (**figure 2.1**)

FIGURE 2.1   Portrait of Isaac Newton (1689), whose work formed the basis of classical mechanics.

In **static analysis**, you may have done things like summing forces at a joint or solving for **reaction forces** in a beam. This assumes that the material that you are loading can handle the load (without failure or extreme deformations). Unfortunately, this is not always the case. In fact, all materials, when subjected to a force or load, will **deform** a little bit (even if it is not visible). These small deformations create significant **internal stress** in the material, which enables the material to "push back" and balance the applied load. The relationship between the amount of deformation and stress in the material can be described by a material's **constitutive law** and analyzed using **mechanical principles**, which is the focus of this chapter.

Mechanics of materials is a subject that deals with the behavior of objects withstanding **stresses** and **strains** (which can be thought of as material equivalents for forces and deformations, respectively). The methods employed to predict the response of materials under loading and susceptibility to various failure modes may take into account various properties other than material yield strength and ultimate strength (e.g., load history, geometry). The mechanical properties and mechanical responses of materials allow us to *objectively (and quantitatively) compare materials for engineering design*. In other words, we can say things like "steel would be better in this structure" *precisely* because we know the *mechanical behavior* of steel. The two key mechanical responses that we consider in civil engineering include the following:

- **Strength:** In practice, the ultimate strength of the materials or maximum load that will dictate design loads. Strength requirements also encompass how the material

---

2   In civil engineering, we typically equate equilibrium with the *static condition only*; that is, we don't like our structures moving with constant velocity.

fails, including brittle (usually catastrophic), ductile (or plastic), or functional (such as fatigue) failure. Understanding and preventing failure is required to design safe structures and systems.

- **Deformation:** Understanding of how materials will deform under external load and the maximum extension of a material and other deformation limits (such as rotation). Typically, deflections are limited to meet serviceability requirements. For example, a footbridge swaying in the wind may be perfectly safe from failure, but people may be nervous to cross if it does not provide stable footing.

Engineering designs must typically satisfy both strength (dealing with *safety*) and deformation (dealing with *serviceability*) requirements, and thus we need to know how to work with the **mechanical properties** of materials with these two concepts in mind.

II. Elastic Behavior and Material "Springs"

This section outlines the basic mechanics of **elastic response**—a physical phenomenon that materials often (but do not always) exhibit. An elastic material is one that deforms immediately upon loading, maintains a constant deformation as long as the load is held constant, and returns immediately to its original undeformed shape when the load is removed. Simply put, it has a "spring-like" response. This section will also introduce two essential concepts in the mechanics of materials: **stress** (σ) and **strain** (ε).

A "Spring" Perspective: Hooke's Law

Since we want to consider how a material deforms owing to applied load, we first consider the simplest case of load and resulting deformation—**linear springs**—as depicted in **figure 2.2**.

Specifically, we consider springs that behave according to **Hooke's law**—a principle of physics that states that the force, F, needed to extend or compress a spring by some distance (a displacement), δ, is proportional to that distance, by some constant, k. That is,

$$F \propto \delta,$$

or

$$F = k\delta,$$

FIGURE 2.2   Depiction of a simple model linear spring.

where k is a constant characteristic of the spring, its **stiffness**. Hooke's law is named after the seventeenth-century

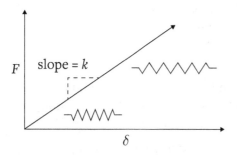

FIGURE 2.3   A linear force *versus* displacement relation for a spring: Hooke's law.

British physicist Robert Hooke.[3] Plotting force versus displacement, this relation looks like a straight line (**figure 2.3**).

Simply put, there is a *linear* relation between the force, F, and the resulting displacement of a spring, δ (see **figure 2.3**). If a spring requires 5 pounds of force to displace it 1 inch, then 10 pounds of force will displace the same spring 2 inches. The slope of the plot is the "stiffness," which is in units of force per length (e.g., Newtons/meter or pounds/inch). For the previous simple example, the stiffness is a simple 5 pounds/inch. Hooke's equation in fact holds (to some extent) in many engineering situations where an elastic body is deformed, such as wind blowing on a tall building, the bowing of a roof owing to snow loads, or the deflection of a suspension bridge. An elastic body or material for which this equation can be assumed is said to be **linear elastic** or "**Hookean**."

Like springs, materials also deform in proportion to the applied load, but unlike a single spring, we want to relate the displacement to the force using analogous normalized parameters strain (ε) and stress (σ), so it holds in all cases. Specifically, if we know force is proportional to displacement, or

$$F \propto \delta,$$

can we assume

$$\sigma \propto \varepsilon?$$

That is, for a given material, is the stress proportional to the strain? It turns out that it is, but we have to look a little closer at what we mean by **stress** and **strain**. Like springs, we first turn to the simplest case: materials subject to tension.

Materials Subject to Tension

Perhaps the most natural test of a material's mechanical properties is the **tension test**, in which a strip or cylinder of the material, having length L and initial cross-sectional area A_0, is anchored at one end and subjected to an external axial load P (a load acting along the specimen's long axis—i.e., in the direction of L) at the other (see **figure 2.4**).

As the load is increased gradually, the axial deformation of the loaded end will increase also (i.e., the specimen *stretches* or *elongates*). Eventually, the test specimen breaks. As engineers, we naturally want to understand such matters as how the deformation is related

---

3   Hooke first stated this law in 1660 as a Latin anagram whose solution he published in 1678 as "*Ut tensio, sic vis*," literally translated as "as the extension, so the force"; or, a more common meaning is "the extension is proportional to the force."

to the load and what ultimate fracture load we might expect in specimens of different sizes.

Normalizing the Load through Stress

Clearly, the amount of load required to deform (and ultimately fail) a specimen is related to the size of the specimen. Consider the following: you can easily manipulate a steel paper clip with your bare hands, but bending a steel beam would be difficult. Yet both are made of the same material (steel). From another perspective, can we say if 1,000 pounds is a large load? Perhaps for an office chair, 1,000 pounds is much too great, but for the steel beams found in a building, 1,000 pounds is not very significant at all.

Therefore, maximum loads (or forces) are not a good metric to compare strengths across materials because of

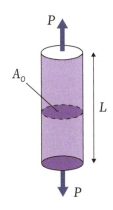

FIGURE 2.4   Typical cylindrical material specimen of length, L, subject to direct tension via a load, P, applied to a surface of known initial cross-sectional area, A_0.

differences in size. One approach is to normalize the forces with area, removing the size dependence. This idea was not always obvious. Ask yourself, "Does a large amount of material necessarily behave proportionally to a small amount? Are there cases when twice the material is not twice as strong?" Indeed, one of the keys to our understanding of material mechanical properties is the realization that the strength of a uniaxially loaded specimen is related to the magnitude of its cross-sectional area, A_0 (here, the subscript "0" indicates the *original* cross-sectional area before the load was applied). This can be thought of in terms of the chemical bonds that make up a material: since the material transfers loads via internal reactions across chemical bonds, more cross-sectional area means more bonds are necessary to be loaded, and thus more load is required to stretch a material (this concept is discussed more in **chapter** 4). It turns out that we can normalize the load per area and get a pretty consistent picture of the load required to stretch/deform a material, *regardless of the size*. With this in mind, a key question arises: "Why doesn't increasing the length of the material increase strength?"

This metric—applied load divided by area—represents the "intensity" of the internal reactions within a material and is called **stress**. When the applied force, P, is perpendicular to the cross-sectional area or plane on which it acts, the stress is called **normal stress**, which is commonly represented by "σ."

When reporting the strength of materials loaded in tension, it is customary to account for the effect of size (area) by dividing the breaking load by the original cross-sectional area, or

$$\sigma_{fail} = \frac{P_{fail}}{A_0},$$

where σ_{fail} is the ultimate tensile stress, often abbreviated as UTS, P_{fail} is the load at fracture, and A_0 is the original cross-sectional area.[4] Since the stress is defined as a load *per unit area*, the units are as follows:

$$\text{Stress} = \frac{\text{Force}}{\text{Length} \times \text{Length}} = \frac{F}{L^2},$$

$$\text{SI Units}: \ \frac{N}{m^2} = \text{Pa},$$

$$\text{US Units}: \ \frac{lb}{in^2} = \text{psi},$$

$$\text{Other US Units}: \ \frac{kip}{in^2} = \text{ksi},$$

where 1 kip = 1 kilopound = 1,000 pounds.

In terms of the **tensile test**, we can use the same relation to define the stress at any load, P, rather than just the failure load, P_{fail}. Thus, the normal stress at a particular load is given by the following equation:

$$\sigma(P) = \frac{P}{A_0}.$$

The tensile stress—the force per unit area acting on a plane transverse to the applied load—is a *fundamental measure of the internal forces within the material*. Much of mechanics of materials is concerned with elaborating this concept to include higher orders of dimensionality, working out methods of determining the stress for various geometries and loading conditions, and predicting what the material's response to the stress will be. **Example 2.1** provides a practice problem for this case.

Normalizing the Deformation through Strain

Like stress, the value of deformation can range from very little to a lot. A 1-inch deflection of a short column may be a relatively large deformation, but a 1-inch sway of the Golden Gate Bridge is not too concerning. What defines a "large" deformation?

---

4   Some materials exhibit substantial reductions in cross-sectional area as they are stretched (to be discussed later in the chapter); using the original rather than final area gives the so-called **engineering** strength.

EXAMPLE 2.1: STRESS ON A SPECIMEN

As a very simple case, let's say we wish to use a steel rod circular in cross-sectional shape to support a load of 5,000 pounds, given the UTS of steel is 60 ksi.

a. *What should the rod diameter be?*

b. *What should the rod length be?*

SOLUTION

a. From $\sigma_{fail} = \dfrac{P_{fail}}{A_0}$, we can rearrange for the necessary area:

$$A_0 = \frac{P_{fail}}{\sigma_{fail}} = \frac{5{,}000 \text{ lbs}}{60{,}000 \dfrac{\text{lbs}}{\text{in}^2}} = 0.0833 \text{ in}^2.$$

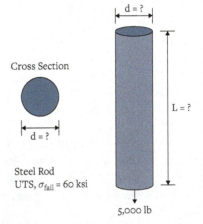

Cross Section

Steel Rod
UTS, σ_{fail} = 60 ksi

5,000 lb

Figure 2.5   Steel rod subjected to tensile loading.

Then we can simply solve for the necessary diameter:

$$d = \sqrt{\frac{4A_0}{\pi}} = 0.326 \text{ in.}$$

Thus, the necessary diameter is at least 0.326 inches or, rounding, perhaps one-third of an inch.

The length of the rod does not matter. It can be 1 inch, 1 foot, or 10 feet long, and it would be able to take the load (in direct tension).

Displacement measures (lengths) are *not* a good metric to compare deformations across materials because of differences in size. One approach is to normalize the deformation with length,[5] removing the size dependence. Like consideration of the original cross-sectional area, A_0, with stress, deformation can be viewed in terms of the **original length** of the system, denoted here as L_0. In effect, strain tells us a proportional or percentage change in length (either increasing or decreasing) and can be considered a measure of the "**intensity of deformation**" (see **figure 2.6**).

---

5   Similar to normalizing *force* by *area* to attain *stress*.

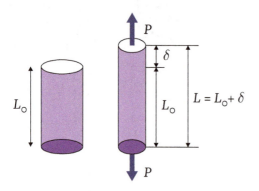

FIGURE 2.6 Schematic of material specimen subject to deformation *via* a tensile load, *P*. The initial length, *L$_0$*, increases by a finite deformation, δ, such that the new length, *L*, can be defined as $L = L_0 + \delta$.

Since strain is defined as a ratio of deformation, δ, to original length, L_0, the units are as follows:

$$\text{Strain } (\varepsilon) = \frac{\text{Length}}{\text{Length}} = \frac{L}{L} = \text{no units,}$$

$$\text{SI Units}: \frac{m}{m} = \text{no units,}$$

$$\text{US Units}: \frac{in}{in} = \text{no units.}$$

One question to consider is, "Does it matter if you use feet or inches when calculating strain?"

Clearly, as shown earlier, the units end up canceling, and strain is considered a **unitless** measure. However, care should be taken to ensure that the units are consistent between lengths—one cannot measure strain in inches/feet!

If the deformation and the original length, L_0, are in the same direction, it is known as linear or axial strain, defined as

$$\varepsilon = \frac{\text{change in length}}{\text{original length}} = \frac{\text{deformation}}{\text{original length}} = \frac{\Delta L}{L_0} = \frac{L - L_0}{L_0} = \frac{\delta}{L_0}.$$

The most useful version is the latter, or

$$\varepsilon = \frac{\delta}{L_0}.$$

Also, many times, we wish to quickly solve for the deformed length, from

$$\varepsilon = \frac{L - L_0}{L_0},$$

which we can easily solve for the deformed length, *L*:

$$L = L_0(1 + \varepsilon).$$

Note that the term in parenthesis "$1 + \varepsilon$" is sometimes defined as the **material stretch**, λ, or

$$\lambda = 1 + \varepsilon.$$

The stretch makes it easy to determine if the material elongates $(\lambda > 1)$ or contracts $(\lambda > 1)$ and is particularly useful when dealing with large deformations. However, most civil engineering applications focus more on the concept of strain as a percent change in length, and it is usually relatively small.

Compression

Many materials are tested in **compression**—that is, we apply a load into the material in a *squeezing* fashion, rather than stretching. Think *pushing* as opposed to *pulling*. Typically, we want to test materials in compression if they are susceptible to cracking, such as concrete, asphalt, and wood. *Why would cracking be critical for tensile loading and not compressive loading?* Subject to compression, any small cracks or voids in the material close, whereas under tension, cracks tend to open. Compression is also used for more brittle materials or when connections are difficult to make. However, it is also common to test all materials in *both* tension and compression for a comparison of behaviors.

Like tension, compressive stress can be calculated as

$$\sigma(P) = \frac{P}{A_0},$$

where, this time, P is a *compressive* load and A_0 is the original area of the material normal to the compressive load (see **figure 2.7**).

In some texts, compressive stress is indicated by a negative sign (which makes some analysis/calculations more convenient); thus, you may see a notation indicating −40 MPa to indicate compressive stress. This is due to the fact that compression loads result in compressive strains that *reduce length* and are negative by definition. Likewise, tension loads result in tensile strains that *increase length* and are positive by definition. This must be the case, which we can see using energetic arguments in the next section.

If you don't use +/− notation to differentiate tensile from compressive stresses, it is good practice to explicitly state tension or compression to avoid confusion/ambiguity. That is,

+30 MPa or 30 MPa [tension]

−40 MPa or 40 MPa [compression].

Note that the sign is important when adding stresses, such that

30 MPa − 40 MPa = −10 MPa = 10 MPa [compression]

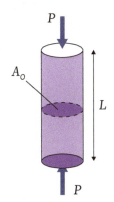

FIGURE 2.7 Typical cylindrical material specimen of length, L, subject to direct compression *via* a load, P, applied to a surface of known cross-sectional area, A_0.

is equivalent to

$$30 \text{ MPa [tension]} + 40 \text{ MPa [compression]} = 10 \text{ MPa [compression]}.$$

Stress, Strain, and Energy

We have stated that "stress" is the "intensity" of an internal reaction, but what exactly does that mean? We know that stretching (or compressing) a spring causes a change in potential or stored energy in the spring.[6] Thus, we can view stress in terms of **energy**.

We note that the discussion that follows is meant to illustrate the relationship between stress, strain, and energy from a *conceptual* approach; it is *not* mathematically rigorous. The connection between stress and energy is formally described by the field of **thermodynamics** and is beyond the scope of this text.

First, we consider some simple concepts of work and energy to derive the energy associated with stress and strain. Consider a force, *F*. What is the **work**, *W*, done by *F*? From physics, we know

$$\text{Work} = (\text{Force})(\text{ Distance}),$$

which also implies no work is done (or can be done) unless the force acts over a little bit of distance. And since we know work is just a form of energy, it follows that

$$\text{Energy} = (\text{Force})(\text{ Distance}).$$

So, we can say that energy *can be* the result of a force acting over a distance. Great! But we still haven't considered stress or strain. We have also said that forces and lengths are *not* good metrics for materials. We do know, however, that

$$\text{Stress} = \frac{\text{Force}}{\text{Original Area}}.$$

And with some rearranging, we get

$$\text{Force} = (\text{Stress})(\text{Original Area}).$$

We can substitute this relation into our equation for energy, arriving at

$$\text{Energy} = (\text{Stress})(\text{Original Area})(\text{Distance}).$$

---

6   If we release the load on a spring, it returns to its original shape, thus releasing the energy. This release of energy is the "potential" energy stored in the spring, which can be made to do **work**.

Now we see that *energy* is a result of *stress*, acting over an *area*, acting over a *distance*. This is still a little vague, so we introduce *strain*. We know

$$\text{Strain} = \frac{\text{Deformation}}{\text{Original Length}}.$$

Now we can consider that the deformation in strain is really equivalent to the distance in which the stress (or force) acts over, and we can rearrange that to arrive at

$$\text{Distance} = \text{Deformation} = (\text{Strain})(\text{Original Length}).$$

Now we are getting somewhere! Substituting this result into our *energy equation*, we get

$$\text{Energy} = (\text{Stress})(\text{Original Area})(\text{Strain})(\text{Original Length}).$$

We now see that the *energy* is a result of the *stress* acting over a *strain*, over a material *area* with some original *length*. Even better, if we know the cross-sectional area of the material and its length, we actually know the *volume* of the material, where

$$\text{Volume} = (\text{Original Area})(\text{Original Length}).$$

Then our relation simplifies to

$$\text{Energy} = (\text{Stress})(\text{Strain})(\text{Volume}).$$

From the aforementioned, we can rearrange slightly and attain the following:

$$\frac{\text{Energy}}{\text{Volume}} = (\text{Stress})(\text{Strain}).$$

We see that the stress and strain are related to an energy *normalized by the volume* in which that stress and strain acts. Typically, when we normalize by a volume, we refer to **density**, and that is exactly the case here. Stress and strain are related to the **strain energy density**, typically denoted as "u." To be a little more technical, it is the change in strain energy density, du, that is related to the change in strain, $d\varepsilon$, because of an applied stress, σ, which can be written as a **differential relation**:

$$du = \sigma \cdot d\varepsilon.$$

This relationship may look a little strange if one has not taken a course on differential equations. However, we can rearrange it slightly again:

$$\sigma = \frac{du}{d\varepsilon},$$

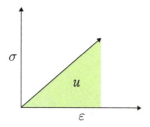

which is a simple derivative (yay for calculus!). That is, stress can be defined as the derivative of strain energy density, *u*, with respect to strain, ε.

From an energy perspective, it is now clear why compressive stresses must be denoted as negative stresses. Considering the earlier derivative, if the strain is negative, then the stress is negative. However, this assumes the change in energy is positive. Is this always the case? As it turns out, yes, straining a system (either stretching it or squeezing it) *always increases the energy* of the system (this is due to the **principle of minimum potential energy**, which will be discussed further in **chapter** 4). With this constraint on strain energy density ($du > 0$), it is obvious from the differential relation that a negative strain implies a negative stress and *vice versa*.

From the differential equation, we can solve for the strain energy density by simply integrating the stress with respect to strain, or

$$u = \int_0^\varepsilon \sigma(\varepsilon)d\varepsilon.$$

This calculation is typically a numerical process, as there are no exact functions, $\sigma(\varepsilon)$, that materials must follow. However, for a linear-elastic material (where stress is proportional to strain—we'll get to that later in the chapter), the differential relation reduces to

$$u = \frac{1}{2}\sigma\varepsilon,$$

which relates the strain energy density to stress and strain. Graphically, the strain energy density represents the area under the stress-strain plot (**figure 2.8**).

This is a very powerful concept in terms of analysis and leads to three important concepts:

1. From the earlier relation, we can see that stress is not simply a load or force applied over any area; that would simply be a **pressure**. A "stress" is a measure of the energy density (energy per volume) because of strain (related to the deformation)—it tells us how much *work* needs to be done to deform a body.

2. Without the "deformation" component (i.e., strain), there can be no stress! This can be thought of in terms of simple work, force, and displacement. If a force does not act over a displacement, then there is no work done by that force, as $W = F \cdot \delta$. Similarly, if a body does not deform (no strain), there is no change in strain energy and thus no stress.

3. Considering stress and strain in terms of energy allows the consideration of different types of energy when solving engineering problems. Indeed, energy is the "great equalizer," and it doesn't really matter the source (e.g., kinetic energy (KE), potential energy, strain energy, thermal energy). We can solve complex problems by reducing everything down to energies! See **Example 2.2.**

Poisson's Ratio

When we test specimens under direct normal stress, we see a peculiar thing happen in the specimen dimensions not aligned with the direction of the stress (see **figure 2.9**).

In tension, specimens tend to get "skinnier" or thinner in the middle, a phenomenon known as **necking**. In compression, we witness **bulging** rather than necking—the specimen expands slightly as we squeeze it. In both cases, we can measure the deformation perpendicular to the applied stress. If we normalize by the original length, D_0, in that dimension, we get the transverse strain, $\varepsilon_{transverse}$, where

$$\varepsilon_{transverse} = \frac{\text{change in length}}{\text{original length}} = \frac{\text{deformation}}{\text{original length}} = \frac{\Delta D}{D_0} = \frac{D - D_0}{D_0} = \frac{\delta_{transverse}}{L_{0,transverse}}.$$

Note that when we stretch a specimen, *increasing* length in the axial direction, the transverse strain *decreases* the width. Also, when we compress a specimen, *decreasing* the length in the axial direction, the transverse strain *increases* the width. The *transverse strains are typically opposite in sign to the axial strains.*

The ratio of transverse strain to axial strain is another material-dependent property, known as **Poisson's ratio**, commonly denoted as ν. It is defined as

$$\nu = -\frac{\varepsilon_{transverse}}{\varepsilon_{axial}}.$$

If the loading is in the x-direction, where the orthogonal y-direction and z-direction define the transverse directions (i.e., Cartesian coordinates), and the material is axisymmetric, then the material's Poisson's ratio is commonly defined as

$$\nu = -\frac{\varepsilon_y}{\varepsilon_x} = -\frac{\varepsilon_z}{\varepsilon_x}.$$

Theoretically, Poisson's ratio can range from −1.0 to 0.5. However, for 99 percent of materials, Poisson's ratio falls in the range from 0 to 0.5. Rubbers and plastics have Poisson's ratios approaching 0.5 and demonstrate large transverse strains. Most metals, on the other hand, have ratios on the order of 0.3 (steel, for example, has a Poisson's ratio of

EXAMPLE 2.2: THE FOOTBALL PLAYER'S FOREARM

Football is a rough sport. Combined with the weight of football players and the high speed of the game, injuries are bound to happen. Luckily, bones are only rarely broken thanks to safety equipment.

However, there are a few areas that players leave unprotected; specifically, let's examine the *forearm*.

Here, we will use our knowledge of *strain energy* to see what could cause a football player's forearm bone to break during a game.

SOLUTION

First, we consider the approximate KE of a football player.

Since players come in a variety of sizes, we will choose to model a tight end. Note that only total mass or weight matters, not the player's height.

Let's assume he weighs approximately 265 pounds, which is 120.2 kilograms, and runs 40 yards (36.6 meters) in 4.65 seconds (measured in the combine, of course). This puts his speed at approximately 7.87 meters per second.

Then, the KE can be calculated as follows:

$$\text{KE} = \frac{1}{2}mv^2 = \frac{1}{2}(120.2\text{kg})\left(7.87\frac{\text{m}}{\text{s}}\right)^2 = 3722.4 \text{ J}.$$

Now we assume the player trips and falls (instantaneously, as to not lose any speed or KE) directly onto his forearm. Even more unluckily, for this problem, we assume that all of the energy from a fall is taken by the bone in his forearm (a worst-case scenario). We also idealize the dimensions of his forearm as a cylinder with a diameter of 5 centimeters and a length of 25 centimeters (remember, we're only considering the forearm bone, not his entire arm). Finally, we assume the bone strains no more than 5% (a reasonable amount for bone). Then we can equate the energy to the strain energy:

$$3722.4 \text{ J} = \frac{1}{2}\sigma\varepsilon V.$$

Note that, to get the total energy, we multiply the energy density by the volume, *V*. Plugging in the numbers we know, we can rearrange and solve for stress:

$$\sigma = \frac{2(3722.4 \text{ J})}{\varepsilon V} = \frac{2(3722.4 \text{ J})}{(0.05)\left(\frac{\pi}{4}\right)(0.05\text{m})^2(0.25\text{m})} = 303{,}328{,}440 \text{ Pa}.$$

Thus, the stress in the tight end's forearm could have potentially reached about **300 MPa**.

The ultimate compressive strength of bone ranges from about 150 MPa to 200 MPa (bone, unlike steel, depends on how healthy a person is).

Regardless, the stress owing to the player's KE at top speed is sufficient to break his arm. *Ouch!*

0.27 to 0.30). More brittle materials, such as concrete and ceramics, have Poisson's ratios on the order of 0.15 to 0.20.

Consideration of the Poisson's ratios is critical when combining materials (e.g., composites) to ensure that they are compatible when deformed. An interesting example of the importance of Poisson's ratio in everyday occurrence is the cork of a wine bottle. Rubber, with a Poisson's ratio approaching 0.5, would expand when compressed into the neck of the bottle and would either jam or break the bottle neck. Cork, on the other hand, has a Poisson's ratio approaching zero and can easily be inserted and removed with predictable forces.

An interesting case also occurs when Poisson's ratio is less than zero (so-called **auxetic** materials). In effect, in the direction of the load under tension, stretching occurs, but in the transverse direction, stretching *also* occurs. These interesting materials significantly increase their volume when subject to tension and have applications in sealed fittings (e.g., straining a seal in one direction will effectively tighten the seal in the other directions).

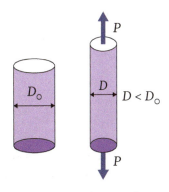

FIGURE 2.9   Schematic of material specimen subject to deformation *via* a load, *P*. The initial diameter, D_o, decreases to a smaller diameter, *D*, transverse to the direction of load.

Change in Volume and Volumetric Strain

Since we can use Poisson's ratio to determine changes in dimension in directions not subject to the load, we can then use all the strains (in three dimensions) to compute associated **volume changes**.

Consider a rectangular specimen (see **figure 2.10**, (a)) of material with initial dimensions of

$$L_{0x} \times L_{0y} \times L_{0z}.$$

We can easily calculate the initial volume $V_0 = L_{0x} L_{0y} L_{0z}$.

The material specimen is subject to a load in the x-direction, and we can measure the resulting axial strain, ε_x. If we know the axial strain (in the x-direction), then we can calculate the transverse strains (in the y and z directions) using Poisson's ratio, ν where

$$\varepsilon_y = \varepsilon_z = -\nu\varepsilon_x.$$

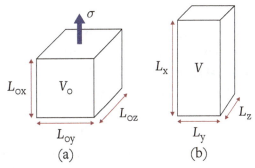

FIGURE 2.10   Volume change in a material specimen subject to tensile stress. (a) Initial material geometry and volume; (b) deformed material geometry and volume.

Since we know all the strains, we can now easily calculate the new dimensions of the specimen, using $L = L_0(1+\varepsilon)$:

$$L_x = L_{0x}(1+\varepsilon_x),$$

$$L_y = L_{0y}(1+\varepsilon_y) = L_{0y}(1-\nu\varepsilon_x),$$

$$L_z = L_{0z}(1+\varepsilon_z) = L_{0z}(1-\nu\varepsilon_x).$$

Note that the lengths are all a function of the axial strain, ε_x, and Poisson's ratio, ν.

With these **new dimensions** (see **figure 2.10** (b)), we can calculate the **new volume** of the specimen:

$$V = L_x L_y L_z = L_{0x}(1+\varepsilon_x)L_{0y}(1-\nu\varepsilon_x)L_{0z}(1-\nu\varepsilon_x) = L_{0x}L_{0y}L_{0z}(1+\varepsilon_x)(1-\nu\varepsilon_x)^2.$$

We note that the first part $(L_{0x}L_{0y}L_{0z})$ is just the **initial volume**, V_0, so we substitute that into the earlier expression, attaining the following:

$$V = V_0(1+\varepsilon_x)(1-\nu\varepsilon_x)^2.$$

This is the exact formulation for the new volume, V, but it is still a little "ugly" because of the strain terms. We can expand the expression, where

$$(1+\varepsilon_x)(1-\nu\varepsilon_x)^2 = 1+\varepsilon_x(1-2\nu)+\nu^2\varepsilon_x^2-2\nu\varepsilon_x^2+\nu^2\varepsilon_x^3.$$

Then we can isolate the "higher-order" strain terms (e.g., ε_x^2 and ε_x^3). Since strains are typically really, really small (say, less than 2 percent), if we square them or cube them, the resulting number is even smaller (by orders of magnitude). Thus, we consider these contributions negligible and *drop* them from our equation. This results in

$$(1+\varepsilon_x)(1-\nu\varepsilon_x)^2 \cong 1+\varepsilon_x(1-2\nu).$$

Given a strain of $\varepsilon_x = 1\%$, for example, and $\nu = 0.3$, the left-hand side of the earlier expression equates to 1.00394909 (exact), whereas the right-hand side of the expression equates to 1.004 (approximate)—a difference of 0.00005091.

We can thus substitute the approximation into our equation for volume, attaining

$$V \cong V_0(1+\varepsilon_x(1-2\nu)) = V_0 + V_0\varepsilon_x(1-2\nu).$$

The aforementioned is now an **approximation**, but is fairly accurate for small values of strain. We may also be interested in the *change in volume*, ΔV, where

$$\Delta V = V - V_0 = [V_0 + V_0\varepsilon_x(1-2\nu)] - V_0 = V_0\varepsilon_x(1-2\nu).$$

EXAMPLE 2.3: POISSON'S RATIO AND VOLUME CHANGE

Consider a rectangular block of aluminum, z = 30 mm, y = 60 mm, and x = 90 mm, placed in a load frame and subject to 100 MPa stress in the x-direction.

a. How much load was required?

b. If E = 70 GPa and ν = 0.333, what will be the increase in the x-direction, assuming linear elasticity? In the y-direction? In the z-direction?

c. What will be the change in volume of the block, assuming small strains?

SOLUTION

d. Using the definition of stress as load over area,

$$\sigma = \frac{P}{A_0} \rightarrow P = \sigma A_0 = (100 \times 10^6 \text{ Pa})(0.030\text{m})(0.060\text{m}) = 180{,}000 \text{ N}.$$

The necessary load is 180 kN.

e. Since the load is in the x-direction, we can use Hooke's law:

$$\sigma = E\varepsilon \rightarrow \varepsilon = \frac{\sigma}{E} = \frac{100 \text{ MPa}}{70{,}000 \text{ MPa}} = 0.001429.$$

Then we can calculate the increase in length:

$$\varepsilon_x = \frac{\delta_x}{l_x} \rightarrow \delta_x = \varepsilon_x l_x = 0.001429(90\text{mm}) \cong 0.13 \text{ mm}.$$

For the y and z directions, the strains are calculated through Poisson's ratio, where

$$\varepsilon_y = \varepsilon_z = -\nu\varepsilon_x = -0.333(0.001429) = -0.000476.$$

The negative means there is a decrease in lengths, where

$$\delta_y = \varepsilon_y l_y = -0.000476(60\text{mm}) \cong -0.029 \text{ mm},$$

$$\delta_z = \varepsilon_z l_z = -0.000476(30\text{mm}) \cong -0.014 \text{ mm}.$$

f. Assuming small strains, we can use our derived formula, where

$$\Delta V = V_0 \varepsilon_x (1 - 2\nu) = [(90\text{mm})(60\text{mm})(30\text{mm})](0.001429)(1 - 2(0.333)) \cong 77.32 \text{ mm}^3.$$

Thus, the volume increases by 77.32 mm².

(As an exercise, you can calculate the change in volume exactly by considering the new dimensions of the block owing to the changes in length calculated in part (b). Is there a difference between volume calculations? Is it significant?)

And with this, we can determine the **volumetric strain**, ε_{vol}, where

$$\varepsilon_{vol} = \frac{\Delta V}{V_0} = \frac{V_0 \varepsilon_x (1 - 2\nu)}{V_0},$$

or

$$\varepsilon_{vol} = \frac{\Delta V}{V_0} = \varepsilon_x (1 - 2\nu).$$

This expression is a quick method to determine the change in volume just by knowing the axial strain and Poisson's ratio of a material specimen. An exact version can be formulated based on the earlier equations (without eliminating the higher-order strain terms—you can try it as practice!), but usually results in differences less than 5 percent for small strains.

Back to a "Spring" Perspective: Material Stiffness

We've so far talked about stresses and strains, but not the *relationship* between the two. We know that when we apply a stress to a material, *it will deform*. Even if it is not noticeable.

To determine *how much* it deforms, we can consider **Hooke's law** once again for a simple linear spring, where

$$F = k\delta.$$

We now know, however, that dividing a force by an area over which it acts gives us a stress, so we can divide each side of the spring law by A_0:

$$\frac{F}{A_0} = \frac{k\delta}{A_0}.$$

This results in the equation relating stress to some spring constant, k, the cross-sectional area of the specimen, A_0, and the deformation owing to the load, δ:

$$\frac{F}{A_0} = \sigma = \frac{k\delta}{A_0}.$$

We also now know that the deformation, δ, can be expressed in terms of the strain, ε, if we know the original length of the specimen, L_0, where $\delta = L_0 \varepsilon$. Substituting this relation results in

$$\sigma = \frac{k(L_0 \varepsilon)}{A_0}.$$

So, we have *stress* on one side and *strain* on the other (**figure 2.11**). We can rearrange the other terms together, finding

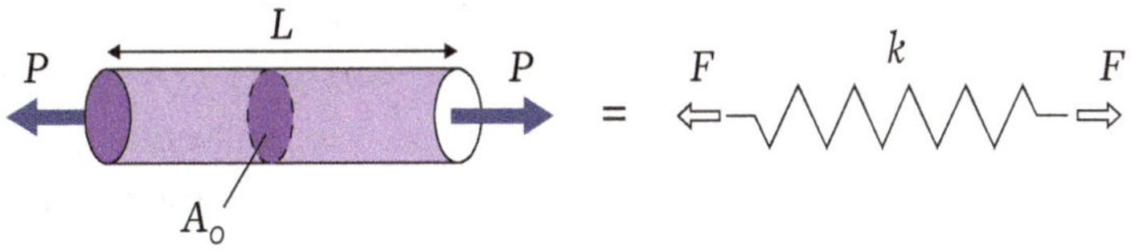

FIGURE 2.11   Materials deform (i.e., strain) when subject to loads (i.e., stress) just as a spring deforms. We can easily "convert" one behavior to another if we know the material "stiffness," *k*, and the geometry of the material specimen (*L, A_o*).

$$\sigma = \left(\frac{kL_0}{A_0} \right) \varepsilon.$$

We see a strange term "kL_0 / A_0" that relates stress to strain, which has components of spring stiffness normalized by *area* and amplified by *length*.

It turns out that like a spring, this "stiffness" term relating the amount of stress (load) with the amount of strain (deformation) is approximately constant for many materials subject to small deformations and can be associated with a material type, and this is the focus of the next section.

III. Relating Stress and Strain through Stiffness

Probably the most important part of mechanical behavior is the **relationship between internal stress and strain**. We can calculate stress by normalizing the load over the area in which it acts, and we can calculate strains by considering the length of the original specimen; we do this specifically to eliminate any dependence on specimen size. Now we want to know, for a given material, *how much strain results when a certain stress level is applied*?

The key point is that since we're using stress versus strain, rather than force versus displacement, we can use the *same* material curve for *different* size specimens. That is, we wish to determine the stiffness of the material, independent of the "size." This stiffness—a factor that relates stress to strain—is typically called the **modulus** of the material.

Elastic Behavior

Elasticity occurs when a material is loaded, **deforms**, and then **recovers** its old shape when unloaded. Think of a spring returning to its original length. In terms of the chemical structure, you can think of elasticity as simply stretching chemical bonds—when a load is released, the bonds simply return to their equilibrium (initial conditions). There is no change in the atomistic or molecular structure. There are two basic types of elastic behaviors: **linear** and **nonlinear**.

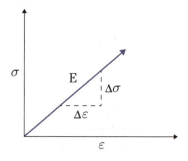

FIGURE 2.12   Plot of linear-elastic, stress-strain relation. The slope of the line is equal to the modulus of elasticity, "E."

Linear Elasticity

Linear elasticity is a special case of elasticity where the relationship between stress and strain is **proportional**, or **linear**. Since engineers love performing calculations,[7] a linear equation makes design calculations a whole lot easier. You will see.

The linear relation is simply Hooke's law applied to stresses and strains, where

$$\sigma = E\varepsilon.$$

Clearly, Hooke's law is a linear function in the form of $y = mx + b$, where the intercept (b) is simply zero (no stress if strain is zero!) and the slope (e.g., the ratio of stress to strain) is simply "E." This slope has a special definition and is known as the **modulus of elasticity**, or **Young's modulus**, defined as

$$E = \frac{d\sigma}{d\varepsilon} \cong \frac{\Delta\sigma}{\Delta\varepsilon}.$$

The modulus tells us the **stiffness of the material** (i.e., how much *stress* is required to cause a unit of *strain*, or stress per strain). Since strain is unitless, the units of the modulus are the same as the units of stress, or force/area [F/L^2], such as MPa or psi. Also, since strain values are typically small, the modulus is typically an order of magnitude (or more) larger than measured stresses (*E* for steel, for example, is approximately 200 GPa ... *gigapascals!*).

For a pure linear-elastic response (**figure 2.12**), the stress-strain plot looks identical to the force-displacement response of a spring (e.g., as depicted in **figure 2.3**):

We can actually relate the **modulus**, *E*, to a spring stiffness, *k*. Recall that we derived

$$\sigma = \left(\frac{kL_0}{A_0}\right)\varepsilon.$$

Comparing this to $\sigma = E\varepsilon$, we see that

$$E = \frac{kL_0}{A_0}.$$

Or, rearranging, we find

$$k = \frac{EA_0}{L_0}.$$

---

7   "The only love that lasts is unrequited love."—Woody Allen

Thus, we can now easily convert a material specimen to an **equivalent spring** if we know its elastic modulus, cross-sectional area, and length. This becomes useful when we wish to analyze structural members in direct tension and/or compression, which is common in truss structures.

EXAMPLE 2.4: EQUIVALENT SPRING STIFFNESS OF A TRUSS

Given a truss member with a length of 1.2 meters and a cross-sectional area of 6.25(10⁻⁴) m² made of steel with a Young's modulus of 200 GPa, do the following:

> a. *Determine the equivalent spring stiffness*
>
> b. *Determine the displacement if under a load of 1200 kN*
>
> c. *If the length is doubled, what is the resulting displacement?*

SOLUTION

a. From the derived equation,

$$k = \frac{EA_0}{L_0} = \frac{(200 \times 10^9 \text{ Pa})(6.25 \times 10^{-4} \text{ m}^2)}{1.2 \text{ m}} = 104.2 \times 10^6.$$

In other words, if we stretched the truss like a spring, it would take approximately 104 million Newtons (N) to stretch it one meter. This is a *huge* load (about 10,000 metric tons). But also, truss members typically don't stretch very much.

b. Now that we have an equivalent spring stiffness, determining displacements is easy.

From $F = k\delta$ it follows that

$$\delta = \frac{F}{k} = \frac{1{,}200 \times 10^3 \text{ N}}{104.2 \times 10^6 \dfrac{\text{N}}{\text{m}}} = 0.0115 \text{ m} = 11.5 \text{ mm}.$$

c. Rather than recalculating the stiffness, *k*, we just notice that the length, L_0, is in the denominator. Thus if the length of the truss is *doubled*, then the stiffness of the truss is *halved*.

If the stiffness is *halved*, then the displacement is *doubled* (because of the relation $F = k\delta$). Thus,

$$\delta = 23 \text{ mm}.$$

(As an extra exercise, confirm this with a full calculation).

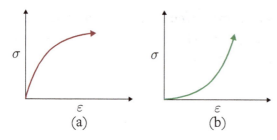

FIGURE 2.13 (a) Plot of nonlinear elastic softening stress-strain relation. The slope of the line decreases as the stress/strain increases. (b): Plot of nonlinear elastic stiffening stress-strain relation. The slope of the line increases as the stress/strain increases.

Nonlinear Elasticity

Nonlinear elasticity essentially covers all the cases in which deformation is recovered, but the *relationship is no longer linear* (it does not follow Hooke's law).

What could the relationship be? It depends on the material.

One common case is a yielding type behavior, where the **material softens** as it is stretched (see **figure 2.13a**).

Another common case is so-called **hyperelastic** stiffening in which the opposite happens (common among rubbers/polymers), and the **material stiffens** as it is stretched (see **figure 2.13b**).

Here we are saying that the stiffness (slope) varies with the stress level and cannot be "defined" as a single number, such as "E." The "stiffness" can, however, be defined at a particular stress level (e.g., the local slope, $\Delta\sigma/\Delta\varepsilon$ or tangent slope, $d\sigma/d\varepsilon$).

Lastly, it is not uncommon for the loading path and the unloading path to take a different shape. This is known as the stress-strain **hysteresis** (see **figure 2.14**).

If you consider the area under the stress-strain curve as the energy, then it can be easily shown that the *area in the loading phase is greater than the area in the unloading phase* (see **figure 2.15**).

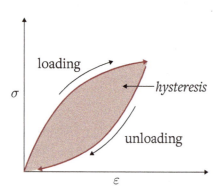

FIGURE 2.14 Plot of nonlinear elastic stress-strain relation with variation between loading and unloading paths. The difference between the paths is known as the *hysteresis.*

Thus, the hysteresis depicts a **change in energy** or energy *loss*. Essentially, during a loading/unloading cycle, the hysteresis indicates how much energy is lost in the process (such as through heat generation). Instead of getting back *all* of the elastic energy we put into a system, we lose some to the surrounding environment.

Can you think of cases where this may be beneficial for engineering design?

Plastic or Inelastic Behavior

Plastic or inelastic behavior is defined when some deformation is permanent upon unloading; that is, when you release a load or stress, the material does not return to its original shape and/or dimensions.

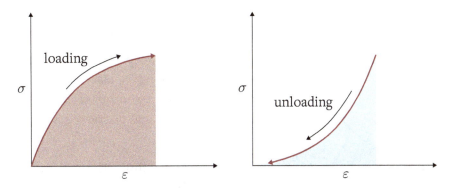

FIGURE 2.15 Separating the strain energy contributions (area under the curve) of loading and unloading phases of a nonlinear elastic material with a hysteresis.

This type of behavior occurs after some **yielding** has occurred in material. Up to the **yield point**, the material still displays an *elastic* response; beyond the yield, permanent deformation occurs (see **figure 2.16**).

From a chemical perspective, plastic behavior occurs when there is a *change in molecular structure*, such as dislocations, molecular unfolding, or the breaking/formation of chemical bonds. Since the chemical structure is no longer in the original form, the *new equilibrium structure* differs from the original.

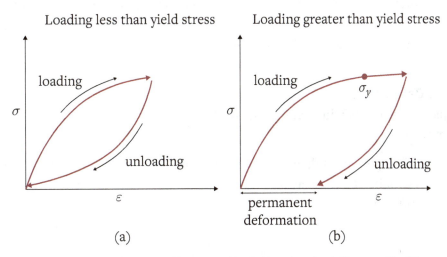

FIGURE 2.16 Onset of plasticity. (a) A material loaded less than the yield stress will still have an elastic response (zero deformation, or $\varepsilon = 0$) when the load is removed; (b) beyond the yield stress (σ_y), plastic behavior occurs, and there is permanent deformation ($\varepsilon \neq 0$), even when the load is removed.

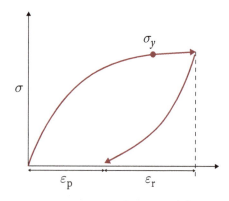

Plastic and recovered strain(s). Upon stress, a material is deformed to a total strain, ε_{total}. When the load is released, some strain may be recovered (ε_r), while the remaining is defined as the plastic strain (ε_p).

As most materials respond with some combination of linear elastic, nonlinear elastic, and plastic behavior under load (depending on the magnitude of the stress), describing the plastic behavior requires the definition of a few other terms to differentiate between elastic and plastic regimes, linear and nonlinear behavior.

- **Proportional limit:** The maximum stress up to which σ-ε ratio is constant (i.e., linear). The proportional limit defines the extent of *linear*-elastic behavior.
- **Elastic limit:** The maximum stress that can be applied to a material and still get all strain recovered—that is, the material returns to its original shape and/or dimensions when the load is released. The elastic limit defines the extent of *elastic* behavior.
- **Plastic regime and recovered strain:** Stress/strain beyond elastic limit—the material can still acquire (some) load, but there is permanent, irrecoverable deformation (known as plastic strain, ε_p). Depending on the material, there is some strain recovered (ε_r). The total strain can be defined as $\varepsilon_{total} = \varepsilon_p + \varepsilon_r$ (see **figure 2.17**).
- **Fracture/ultimate failure:** Defined by a crack or separation of material; the material is broken and cannot sustain any load.

Modulus Definitions

As we have hinted, depending on the use, the "modulus" or "stiffness" can be defined in different ways.

1. **Initial tangent modulus:** A linear fit of the initial stress-strain response (starting at the origin, capturing as many initial points as possible). Mathematically, it can be written as

$$E_{initial}(0) = \frac{d\sigma}{d\varepsilon}\bigg|_{\varepsilon=0.}$$

Numerically, one would limit the fitting to a very small strain range, perhaps less than 1 percent to 2 percent or 0.1 percent to 0.2 percent, depending on the material type.

2. **Tangent modulus:** Derivative of the stress-strain curve at a given stress/strain point. It gives the instantaneous or local stiffness:

$$E_{tanget}(\varepsilon) = \frac{d\sigma}{d\varepsilon}\bigg|_{\varepsilon}.$$

Again, numerically, if a functional form of the stress to strain relationship is unknown, one could define a small range in the area of interest to numerically fit a tangent line.

3. **Secant modulus:** The slope that connects origin and any point on curve:

$$E_{secant}(\sigma,\varepsilon) = \frac{\sigma}{\varepsilon}.$$

While a simple, and often poor, approximation for stiffness, the secant modulus is useful, as it gives the necessary stress to attain a given strain. Thus, it is sometimes referred to as the "apparent" or "effective" stiffness of material.

4. **Chord modulus:** The stiffness between any two points on the stress-strain curve. A close approximation to the average stiffness along a segment:

$$E_{chord}(\sigma_1,\sigma_2,\varepsilon_1,\varepsilon_2) = \frac{\Delta\sigma}{\Delta\varepsilon} = \frac{\sigma_2 - \sigma_1}{\varepsilon_2 - \varepsilon_1}.$$

Of note is that as the stress and strain values used to compute the chord modulus approach each other, the chord modulus approaches the tangent modulus.

IV. Yield, Ultimate Strength, and Other Stress-Strain Properties

The stress-strain curve gives us much more information than the stiffness of the material. It also tells us a lot about how material behaves under load and indicates how the material may fail.

Yield Point/Strength

When you strain a material, the internal stress subsequently increases. If you continue to strain the same materials, sometimes there comes a point where strain increases and either the stress does not increase, or the stress actually decreases—this is called the yield point (**figure 2.18** (a)). Mathematically, it is the point at which the slope of the stress-strain curve equals zero. Commonly, the yield point does not equal the maximum stress that the material can maintain; materials do not usually fail when they yield.

The **yield strength** (**figure 2.18** (b)) is defined as the stress state at which the material starts to "give"—that is, it takes a little less stress to produce a little more strain. This can be visualized as a decrease in slope in the stress-strain curve (decrease in slope = decrease

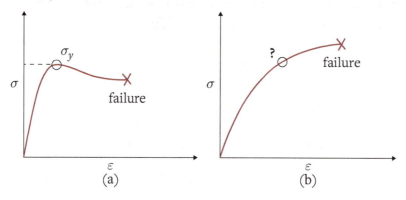

FIGURE 2.18 Material yielding occurs when there is an intrinsic softening of a material under stress. Plot (a) depicts a material with a clear *yield point*—a local "peak" in the stress-strain curve where the slope equals zero. Beyond this strain, the material is said to have "yielded." Plot (b) also depicts a yielded material, but without a clear yield point (i.e., no "peak"). The *yield strength* can be estimated by offsetting the initial tangent stiffness.

in stiffness = softening). The yield strength provides an estimate where plastic deformation will start to occur. Since plastic deformation is permanent, we typically like to engineer the allowable stresses applied to materials below their respective yield strengths. Depending on the material, there may not be an obvious "point" at which yielding occurs (the entire stress-strain curve may be somewhat nonlinear and continuously "softening"). However, we can estimate the yield strength. The yield strength can be approximated using the standard **offset method** (see **figure 2.19**):

1.   Determine the initial tangent modulus (either graphically or by calculation).
2.   Pick a small strain offset, $\Delta\varepsilon$ (0.1 percent to 0.3 percent; commonly 0.2 percent), and draw a line from $\Delta\varepsilon$ parallel to the initial tangent modulus.
3.   Intersection with the curve gives the approximate yield strength.

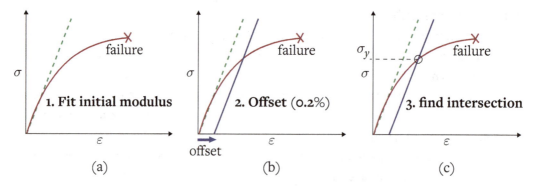

FIGURE 2.19 Schematic of the offset method. The yield strength, σ_y, is defined at the intersection of an offset line (commonly at $\varepsilon = +0.2\%$), parallel to the initial tangent modulus, and the stress-strain curve.

Other Stress-Strain Properties

Commonly, materials have a complex stress-strain behavior, combining elements of elasticity, plasticity, and yielding, among others. A few other common properties (both quantitative and qualitative) are described in **table 2.1.**

Figure 2.20 depicts a stress-strain plot representative of a metal, such as steel, to illustrate some concepts.

TABLE 2.1 COMMON STRESS-STRAIN PROPERTIES

Property	Description
Ultimate strength	The maximum stress attained by a material at any point before fracture (e.g., the highest point on $\sigma - \varepsilon$ curve). If a material specimen was loaded by incremental forces (rather than incremental displacements), then the ultimate strength would reflect the failure load of the specimen.
Ductility	A qualitative measure that describes a material's ability to undergo a lot of strain and/or plastic deformation before fracture. Examples of ductile materials include some plastics, wrought iron, steel, copper.
Brittleness	The tendency of material to fracture without significant deformation; the "significant" definition will depend on the material. Examples of brittle materials include concrete, glass, brick.
Strain hardening	Occurs if a material requires increased stress to strain further (increased stiffness).
Strain softening	Occurs if a material requires less stress to continue straining.
Toughness	A material's ability to build up strain energy; quantitatively, the area under $\sigma - \varepsilon$ curve.
Resilience	The strain energy absorbed up to the elastic limit; a measure of the energy that can be released if a load is removed.

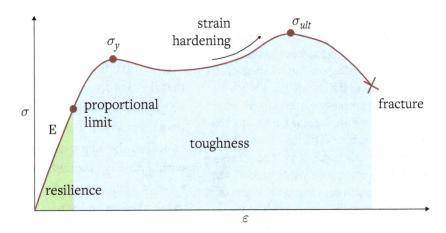

FIGURE 2.20   Representative stress-strain plot of steel depicting some key parameters as defined in table 2.1.

V. Other Loading/Deformation Conditions

Other than directly pulling (tension) or pushing (compression) on a material, there are other ways to apply load and deformation. Here, we outline the basic equations and relations for (a) bending, (b) shear, (c) torsion, and (d) thermal loading.

Bending

Bending occurs when a material is subject to curvature, κ, typically by an applied moment, M. The moment can be thought of as a force couple that acts about an axis of bending. Since there are effectively two forces that act in opposite directions, bending is really a combination of normal stresses consisting of compression on the top and tension on the bottom (for a positive moment), as indicated in figure 2.21.

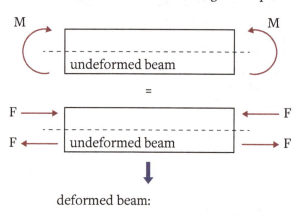

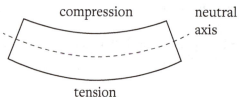

FIGURE 2.21   Bending because of moments (top) can be easily pictured by thinking of the moment as a force couple (middle) that acts about a bending axis. This couple causes the beam to curve (bottom). Since the force couple acts into the beam at the top, there is a compressive stress, and the top section decreases in length. Similarly, since the force couple acts out of the beam at the bottom, there is a tensile stress, and the bottom section increases in length. At some point between the top and bottom, there is no change in length and thus no stress—this is called the neutral axis of the beam. In general, the exact location of the neutral axis is dependent on the beam cross-sectional geometry, as well as the applied load(s).

In **elastic bending** (e.g., a linear stress-strain relationship), the compression/tension stresses are related to the distance from the neutral axis (where stress is zero), y, according to the *flexure formula*

$$\sigma = -\frac{My}{I}.$$

The parameters in this equation are defined in **table 2.2**. As indicated by the arrow in **figure 2.22**, values of y are positive above the neutral axis in the cross-section and negative below; at the neutral axis, $y = 0$ (there is no stress at the neutral axis).

Recall from earlier in the chapter that the sign convention for stress will be negative for compression and positive for tension. The flexure formula contains the negative sign because with a positive moment and positive value of y (for points above the neutral axis), the normal stress is compressive, or negative (see **figure 2.21**).

TABLE 2.2 ELASTIC BENDING PARAMETER DEFINITIONS

Parameter	Definition	Units
σ	Stress at a particular point in a beam cross-section	F/L^2
M	Moment applied to beam (degree of bending)	F-L
y	Distance of point from "neutral axis" (different for different shapes)	L
I	Area moment of inertia (calculus, see **table 2.3** for examples)	L^4

TABLE 2.3 INERTIA FORMULAS AND NEUTRAL AXIS LOCATIONS FOR COMMON SHAPES

Cross-section shape	Figure	Moment of inertia I	Neutral axis (from Bottom)
Rectangle		$I = \dfrac{1}{12}bh^3$	$c = \dfrac{1}{2}h$
Circle		$I = \dfrac{\pi}{4}r^4$	$c = r$
Triangle		$I = \dfrac{1}{36}bh^3$	$c = \dfrac{1}{3}h$
Trapezoid		$I = \dfrac{d^3(a^2 + 4ab + b^2)}{36(a+b)}$	$c = \dfrac{(2a+b)d}{3(a+b)}$
I-Section		$I = \dfrac{bh^3}{12} - \dfrac{(b-t)d^3}{12}$	$c = \dfrac{1}{2}h$

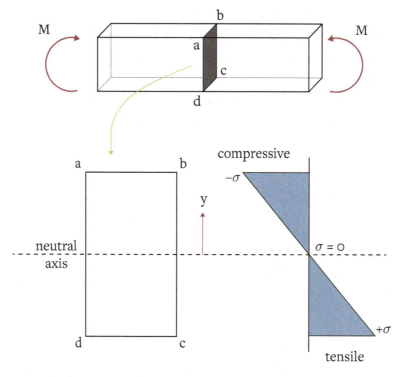

FIGURE 2.22 Stress distribution owing to bending. Given a beam under bending, the stress acts along the beam axis, which is normal to the cross-section (here a rectangle, abcd). By taking this beam "slice," the neutral axis is in the middle (because of the rectangular symmetry) where stress equals zero, and the stress varies from maximum compression at the top to maximum tension at the bottom. Note that y is defined relative to the neutral axis.

We also note that this equation only holds for the cases in which the strain distribution is linear as well (which is why it is known as elastic bending). This equation is more formally derived in higher classes.

Based on the flexure formula, we also recognize that for a given cross-section, the stress distribution is a linear function according to "y," with maximum stress occurring at the top/bottom of the beam (**figure 2.22**).

Strength of material in bending is due to a combination of compression/tension resistance and is more complex than direct compressive or tensile strength. This "ultimate bending stress" is known as the **modulus of rupture**, or "MOR," and is sometimes designated f_r (for concrete, for example), where

$$MOR = f_r = -\frac{M_{max}\,c}{I}.$$

Here, c refers to the maximum distance from the neutral axis to top or bottom edge of the cross-section, corresponding to the point of *maximum stress* by magnitude on the cross-section. Note that the c will be positive or negative depending on whether the maximum distance from the neutral axis to the edge of the beam is above or below the neutral axis (in the positive or negative y-direction, respectively).

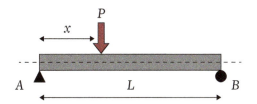

FIGURE 2.23   A simply supported beam with a load applied at an arbitrary distance "x" from the pin support A.

The determination of the MOR is (slightly) more complicated than a compressive or tensile test, as the location of the applied load affects the maximum moment. For example, in the simple case that follows (**figure 2.23**), the maximum moment depends on the location of the load, P.

For this problem, it can be shown that

$$M_{max} = Px\left(1 - \frac{x}{L}\right).$$

This makes it difficult to compare test specimens if there are too many variables to specify. Instead, for **flexural** or **bending tests**, standard loading conditions are traditionally implemented.

Three-Point Bending Test

Three-point bending consists of a simply supported beam with a vertical load "P" applied to the exact center of the beam span, as in **figure 2.24**. In this way, the only loads acting on the beam are the load, P, and the two reactions (thus loaded at *three* points).

It can easily be shown that the reaction loads are both equal to half of the applied load, or P/2.

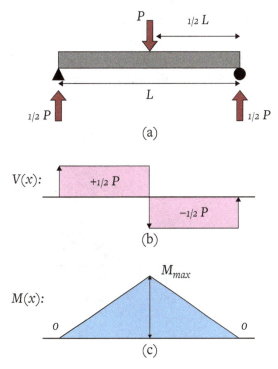

FIGURE 2.24   Three-point bend test: (a) applied loading and reactions, (b) shear diagram, and (c) moment diagram.

Likewise, the maximum moment occurs at the midspan (under the applied load), where

$$M_{max} = \frac{PL}{4}.$$

Of note is that, at that point, there is also a shear load of P/2. As such, failure is a combination of axial stress (because of bending) and shear stress.

Moreover, there is only a single location where the moment is maximum (at the center). If the material specimen fails at another location, it is (sometimes) difficult to determine what the moment at failure was exactly. To rectify these problems of the three-point bending test, there is the alternative four-point bending test.

Four-Point Bending Test

Four-point bending consists of a simply supported beam with a vertical load "P" applied to the beam, but split such that two P/2 loads are applied symmetrically about the center (see **figure 2.25**). The loads acting on the beam now are the two split loads and the two reactions (thus loaded at four points).

Again, it can easily be shown that the reaction loads are both equal to half of the applied load, or P/2. For this case, the maximum moment is *constant* between the loading points:

$$M_{max} = \frac{PL}{6}.$$

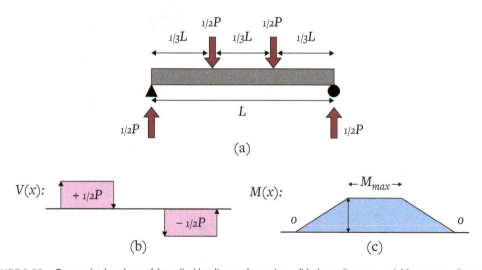

FIGURE 2.25   Four-point bend test: (a) applied loading and reactions, (b) shear diagram, and (c) moment diagram.

EXAMPLE 2.5: BENDING

A square wood dowel with a 2.0-centimeter by 2.0-centimeter cross-section is tested in four-point bending. The supports are located 30 centimeters apart, giving L = 30cm. The load application points are located 10 centimeters apart.

    a. Determine the location of the neutral axis in the cross-section, in the vertical direction.

    b. If an applied force, P = 920 N causes the dowel to fail, determine the modulus of rupture for the wood, in units of kPa.

    c. What is the stress in the wood 0.5 centimeters below the neutral axis at the moment of failure?

SOLUTION

    a. The neutral axis is at the centroid of the cross-sectional area; in this case the cross-section is square, which is doubly symmetric. The neutral axis will therefore be located at the height of the square divided by 2; or **h/2 = 2.0cm/2 = 1.0cm up from the bottom of the dowel** *(or down from the top).*

    b. First, calculate M_{max} from applied load, P = 920 N:

$$M_{max} = \frac{PL}{6} = \frac{920N(0.30m)}{6} = 46N \cdot m = 0.046kN \cdot m.$$

Next, calculate the moment of inertia, I:

$$I = \frac{bh^3}{12} = \frac{0.02m(0.02m)^3}{12} = 1.33 \times 10^{-8} m^4.$$

The maximum distance from the neutral axis to the outer surface of the dowel is

$$c = \pm \frac{h}{2} = \pm \frac{0.02m}{2} = \pm 0.01m \quad \text{(doubly symmetric, same top and bottom).}$$

Calculate the modulus of rupture:

$$MOR = f_r = \frac{M_{max}c}{I} = \frac{0.046kN \cdot m(0.01m)}{1.33 \times 10^{-8} m^4} = 34,600kPa.$$

The modulus of rupture is **f_r = 34,600 kPa**. Since the dowel is symmetric, the stress is the same at the top (compression) and bottom (tension) surfaces.

    c. At 0.5 centimeters below the neutral axis, y = −0.5 centimeters:

$$\sigma = -\frac{My}{I} = \frac{0.046kN \cdot m(-0.005m)}{1.33 \times 10^{-8} m^4} = 17,300kPa \quad \text{(tension).}$$

At 0.5 centimeters below the neutral axis, the stress in the wood dowel, **σ = 17,300kPa** in tension

Also, between the loading points, the *shear load is zero*. Thus, for the four-point test, failure is purely due to **bending stresses**.

Since there is a range between the applied loads in which the maximum moment is constant, as long as failure occurs there, we know the exact moment. Thus four-point bending is important to use when the material is heterogeneous or may have flaws (e.g., concrete, wood).

We typically perform either the three- or four-point bending tests to attain the MOR. If we know the dimensions of the rectangular beam sample (i.e., $b \times d$) we can simplify the aforementioned moment equations to directly calculate the MOR from the applied load, where

$$\text{MOR 3-point test}: \quad f_r = 1.5 \frac{PL}{bd^2}$$

and

$$\text{MOR 4-point test}: \quad f_r = \frac{PL}{bd^2}.$$

Shear Stress

Shear stress encompasses the second-most common type of stress. If you think of normal stress as changes in dimension/length, you can think of shear stress as changes in shape (see **figure 2.26**).[8]

Like normal stress, shear stress is due to a load that acts on a surface. But rather than act perpendicular (normal) to the surface, shear stress occurs when loads act *parallel* to a surface. If we consider a shear load, V, then the shear stress, τ, is defined as

$$\tau = \frac{V}{A_0}.$$

The equation is exactly like the equation for normal stress—the only difference is the direction of the load. As such, the units are equivalent (force/area, psi, MPa, etc.).

Like normal stress, shear stress also results in deformation in the direction of the applied load. Thus, rather than stretching or compressing a specimen, the shear stress effectively skews the surface in a particular direction, called a **distortion**. The ratio of the amount

---

8 Technically, shear deformations can be quantified by changes in the material *angles*.

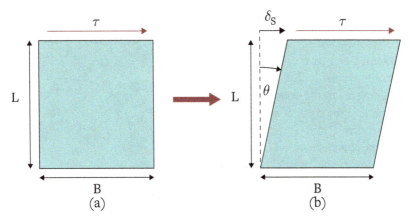

FIGURE 2.26 Deformation owing to shear stress. A shear stress, τ, acts parallel to a surface, resulting in a distortion of the shape. Note that the dimensions L and B are the same between both undeformed (a) and deformed (b) systems. The shear strain, γ, can be defined as the ratio of shear deformation, δ_s, to the specimen length normal to the direction of displacement, L.

the specimen skewed to the length perpendicular to the shearing plane defines the shear strain, or

$$\gamma = \frac{\text{shear deformation}}{\text{original length} \perp \text{to shear stress}} = \frac{\delta_s}{L_0}.$$

Like regular strain, since shear strain is defined as units of length over length; it is a unitless value. We also note that the shear strain can be viewed in terms of two sides of a triangle (of lengths δ_s and L). From this perspective, the shear strain can be defined as the *tangent of the angle of distortion*, where

$$\tan(\theta) = \frac{\delta_s}{L_0}.$$

Moreover, since the strain is typically a small value, we can use the approximation that $\tan\theta \approx \theta$ when $\theta \to 0$, leaving

$$\gamma = \theta = \frac{\delta_s}{L_0}.$$

Of note is that loads that are acting on an angle can be broken into both normal and shear components (based on the load vector). Thus, *a single load can induce both normal and shear stresses.*

EXAMPLE 2.6: SHEAR IN A SHACKLE PIN

A steel shackle is being used on a skidder to drag a cut tree out of the forest for processing into lumber. As the tree is dragged by the skidder, a load of 3,000 pounds is applied to the 0.5-inch diameter cylindrical shackle pin, as shown.

What is the shear stress and shear strain in the shackle pin if the elastic shear modulus, $G = 11.5 \times 10^6$ psi?

SOLUTION

First use a free body diagram of the pin and equilibrium to determine the shear force in the pin:

Shear Force, $V = \dfrac{P}{2} = 1{,}500\text{lb}$ at each end of the pin

Next, determine the cross-sectional area of the pin (the area on which the shear force acts) and calculate the shear stress:

$$A = \frac{\pi}{4}d^2 = \frac{\pi}{4}(0.5\text{in})^2 = 0.196\text{in}^2$$

$$\tau = \frac{V}{A} = \frac{1{,}500\text{lb}}{0.196\text{in}^2} = 7{,}650\text{psi}.$$

Now, calculate the shear strain:

$$\tau = G\gamma \rightarrow \gamma = \frac{\tau}{G} = \frac{7{,}650\text{psi}}{11.5 \times 10^6\,\text{psi}}$$
$$= 6.65 \times 10^{-4}, \text{ or } 0.0665\%.$$

The shackle pin experiences τ = **7,650psi** and γ = **0.0655%**.

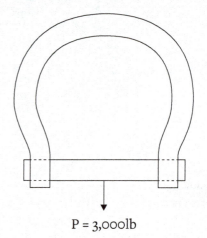

Figure 2.27   Load on skidder shackle pin.

P = 3,000lb

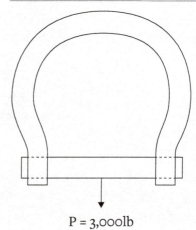

P = 3,000lb

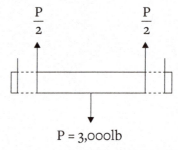

$\dfrac{P}{2}$ $\dfrac{P}{2}$

P = 3,000lb

Figure 2.28   Free body diagram of shackle pin.

Like normal stress, **Hooke's law** can be formulated for shear, relating shear stress to shear strain, where

$$\tau = G\gamma.$$

This requires a definition of the **"shear stiffness"** of the material, which is defined as the **shear modulus**, G. Like Young's modulus, the shear modulus is a linear relation between stress and strain (a slope of the stress-strain plot), where

$$G = \frac{d\tau}{d\gamma} \cong \frac{\Delta\tau}{\Delta\gamma}.$$

Bulk Modulus

Normal stress typically results in stretching (or compressing) a dimension, while shear stress changes a material's shape. What if we want to change the volume of a material sample?

A uniform volume change would involve the simultaneous application of stress in every direction—a loading condition more commonly known as **hydrostatic pressure**. The **bulk modulus**, K, of a substance measures the substance's *resistance to uniform (hydrostatic) compression*. It is defined as the ratio of the infinitesimal pressure increase to the resulting relative decrease of the volume, or

$$K = -V\frac{dP}{dV}.$$

This relation doesn't look much like what we have been seeing, so let us rearrange it a little. First, we put all the volume terms on the same side with the modulus, K:

$$dP = -K\frac{dV}{V}.$$

The "dV" simply means a small change in volume, and the "V" refers to the initial volume (before the change); thus, "dV / V" is just another version of "$\Delta V / V_0$," or volumetric strain. Thus,

$$dP = -K\varepsilon_{vol}.$$

We also note that "dP" simply means an increase (change) in pressure, so we let

$$\Delta P = -K\varepsilon_{vol.}$$

Note that the negative is there because an *increase* in pressure results in a *decrease* in volume. Now the earlier equation looks just like another version of **Hooke's law**, this time relating pressure to volumetric strain through the bulk modulus, K. The stiffer the material, the more pressure it will take to change the volume.

So far, we have laws for *normal stress*,

$$\sigma = E\varepsilon,$$

shear stress,

$$\tau = G\gamma,$$

and *pressure*,

$$\Delta P = -K\varepsilon_{vol}$$

that all take the form of *Hooke's law*, where

$$\text{load} = (\text{stiffness})(\text{deformation}).$$

Of interest is that the parameters that describe stiffness (E, G, and K) are all in units of stress. Moreover, a material that has a high value of one of the parameters typically has high values for all the parameters. Does this make sense? As it turns out, all of these stiffness parameters (or moduli) are *NOT* independent of each other and are also related to Poisson's ratio, ν. These relations are given in **table 2.4**.

Again, without getting into mathematical details, we can see the implications of the theoretical limits of Poisson's ratio (between −1.0 and 0.5). For example, the relation

$$E = 2G(1+)$$

would result in a negative Young's modulus for < -1.0. Similarly, the relation

$$K = \frac{E}{3(1-2)}$$

would imply a negative bulk modulus for > 0.5.

TABLE 2.4 RELATIONSHIP BETWEEN YOUNG'S, SHEAR, AND BULK MODULI

	$f(E,\nu)$	$f(G,\nu)$	$f(K,\nu)$	$f(E,G)$	$f(E,K)$	$f(G,K)$
$E =$	E	$2G(1+\nu)$	$3K(1-2\nu)$	E	E	$\dfrac{9KG}{3K+G}$
$G =$	$\dfrac{E}{2(1+\nu)}$	G	$\dfrac{3K(1-2\nu)}{2(1+\nu)}$	G	$\dfrac{3KE}{9K-E}$	G
$K =$	$\dfrac{E}{3(1-2\nu)}$	$\dfrac{2G(1+\nu)}{3(1-2\nu)}$	K	$\dfrac{EG}{3(3G-E)}$	K	K

It is left to the reader to consider why a negative Young's modulus and negative bulk modulus are not possible.

Torsion

The final type of applied load we will discuss is torsion. Torsion involves the **twisting** of a material when loaded by moments or **torques**. Like a bending moment, a torque causes *rotation about an axis*. For bending, the rotation occurs about an axis perpendicular to the span of the member. For torsion, the *rotation occurs about an axis parallel to the longitudinal span* of the member and thus involves twisting (see **figure 2.29** (b)).

What is a moment or torque? A torque, or twisting moment, T, can be easily envisioned by imagining a screwdriver or corkscrew. Like a bending moment, it can come from a force couple (see **figure 2.29** (a)), where torque is equal to a force couple, P, multiplied by the distance between the couple, d, given by the equation

$$T = P(d).$$

For our purposes, we consider the torsion of circular rods only (more complex shapes are handled in higher-level mechanics classes). We note a few key points:

1. Stress owing to twisting/torsion is a shear stress.
2. Deformation owing to torsion is a rotational deformation.
3. We only consider "pure torsion" (e.g., all slices remain planar and circular).

The common condition is a cylindrical bar/rod with an applied torque on one end and fixed at the other end. The torque will cause the end of the bar to rotate or twist by an angle, ϕ (see **figure 2.30**):

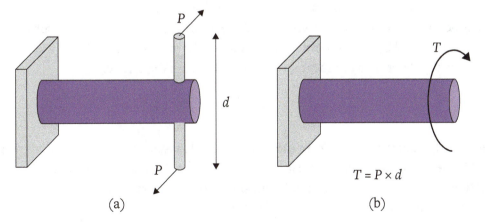

(a)   (b)

FIGURE 2.29   Applied torque at the end of a solid rod either by (a) a force couple or (b) twisting moment.

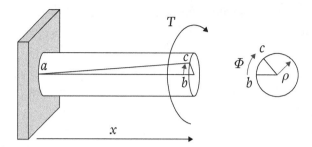

FIGURE 2.30 Applied torque at the end of a solid rod by a twisting moment, T. Upon twisting, the line "ab" is distorted to line "ac." Looking at the circular cross-section, we see that the point that used to be at "b" has rotated to "c." The angle is defined as the angle of twist, ϕ.

From the earlier figure, it is clear that depending on how close we are to the fixed end at "a," the angle of twist decreases (in fact, at "a," or $x = 0$, $\phi = 0$). We can thus define the **rate of twist** as

$$\frac{d\phi}{dx},$$

which indicates how much the angle, ϕ, changes along the bar length, dx. For a bar of total length, L, with a constant rate of twist, this simplifies to ϕ/L.

The **angle of twist** can be related to the shear strain, where the maximum shear strain occurs at the end of the bar:

$$\gamma_{max} = \frac{\phi r}{L},$$

where r is the radius of the bar. This can be easily justified by considering the deformation of the surface of the bar that, when twisted, is displaced by the arc length, $\delta = \phi r$. Substituting this equality into the previous formulation results in the previous definition of shear strain.

We can also determine the shear stress associated with the torque:

$$\tau = \frac{T\rho}{J},$$

where ρ is the **radial distance from the center** (of the twist axis), T is the **applied torque**, and J is the **polar moment of inertia** (for a circular cross-section, $J = \pi r^4 / 2$). From this equation, we easily see that the maximum shear stress occurs at the surface of the rod, where $\rho = r$, and thus

$$\tau_{max} = \frac{Tr}{J}.$$

EXAMPLE 2.7: TORSION IN A SCREW

A screwdriver is being used to install a 5-millimeter diameter, 150-millimeter-long steel screw into a wood stud in order to mount a shelf on a wall. The torque applied to the screw head from the screwdriver is 1.35 N·m when the screw has been screwed into the wall 50 millimeters (100 millimeters extending out of the wall). The elastic shear modulus of the steel screw is G = 79 GPa.

   What is the maximum shear stress, shear strain, and angle of twist (in degrees) in the screw?

SOLUTION

Begin by calculating the maximum shear stress based on the torque for the cylindrical screw:

$$\tau_{max} = \frac{Tr}{J} = \frac{Tr}{\frac{\pi r^4}{2}} = \frac{1.35 N \cdot m (0.005m)}{\frac{\pi (0.005m)^4}{2}} = 6,875,000 Pa = 6.875 MPa$$

Next, calculate the shear strain based on the shear stress:

$$\tau = G\gamma \rightarrow \gamma = \frac{\tau}{G} = \frac{6.875 MPa}{79,000 MPa} = 8.7 \times 10^{-5}$$

Now, calculate the angle of twist, where L = length of screw outside the wall = 100 mm:

$$\gamma_{max} = \frac{\phi r}{L} \rightarrow \phi = \frac{\gamma_{max} L}{r} = \frac{8.7 \times 10^{-5}(0.1m)}{0.005m} = 1.74 \times 10^{-3} \text{radians} \frac{180°}{\pi \text{ radians}} = 0.1°$$

The screw is experiencing τ_{max} = **6.875 MPa**, γ_{max} = **8.7 × 10⁻⁵**, and ϕ_{max} = **0.1°**.

Moreover, we directly calculate twist angles from the torque, substituting Hooke's law for shear ($\tau = G\gamma$), as well as the equation relating shear strain to twist angle, attaining

$$\phi = \frac{TL}{GJ}.$$

Temperature-Induced Stress and Strain

How can we quantify temperature change–induced stresses/strains? If we measure strains with regard to temperature changes, we can fit a **coefficient of thermal expansion**, α, which is in units of strain/degree temperature. Depending on the application, there are two coefficients we can consider:

1. Linear strain version: α_L = amount of linear strain per degree ΔT

$$\alpha_L = \frac{\Delta L}{L} \frac{1}{\Delta T} = \frac{\varepsilon}{\Delta T}$$

2. Volume strain version: α_V = amount of volumetric strain per degree ΔT

$$\alpha_V = \frac{\Delta V}{V} \frac{1}{\Delta T} = \frac{\varepsilon_{vol}}{\Delta T}$$

Both of these expressions are more commonly listed solving for strain, or

$$\varepsilon = \alpha_L \Delta T,$$

$$_{vol} = \alpha_V \Delta T.$$

Assuming isotropic elastic behavior from the relation between strain and volumetric strain, we can easily relate the two thermal expansion coefficients through Poisson's ratio, where

$$\alpha_V = \alpha_L (1 - 2\nu).$$

VI. Concluding Remarks

We started this chapter by introducing the subject of mechanics, broadly defined as a focus on the behavior of physical bodies when subjected to forces or displacements. Now we know a whole set of different stresses (which are like forces) and strains (which are like displacements) that can describe the "mechanical behavior" of materials.

In general, we see the following:

1. Forces and loads lead to stresses.
2. Stresses lead to deformations.
3. Deformation resulting from stress depends on the "stiffness" of the material, described by a set of parameters (e.g., E, G, K).

Quantifying the stress and deformations requires a large analytical tool set. However, there are a number of terms that are material dependent (such as moduli, thermal expansion

EXAMPLE 2.8: MATERIALS CHANGING LENGTH WITH TEMPERATURE

A steel bar is machined at a temperature of 40°F to fit into a space that is 6-feet long. The bar has a diameter of 2.1 inches, E = 30,000,000 psi, and a linear coefficient of thermal expansion of $5(10^{-6})$ $(°F)^{-1}$. On the day of installation, the temperature is 95°F.

What force is required on the bar to keep its length exactly 6 feet during installation? Does the force apply compression or tension to the bar?

SOLUTION

First, we find the change in temperature:

$$\Delta T = 95° - 40° = 55°F.$$

Then we can calculate the strain in the bar:

$$\varepsilon = \alpha_L \Delta T = 5(10^{-6})(°F)^{-1}(55°F) = 0.000275.$$

Using Hooke's law, we calculate the necessary stress to achieve equal strain:

$$\sigma = E\varepsilon = 30(10^6)\text{psi}(0.000275) = 8,250 \text{ psi},$$

find necessary force knowing the cross-sectional area:

$$F = \sigma A_0 = \sigma \left(\frac{\pi d^2}{4}\right) = (8,250\text{psi})\frac{(\pi)(2.1\text{in})^2}{4} = 28,575 \text{ lb},$$

or, in one single calculation:

$$P = E\alpha_L \Delta T \left(\frac{\pi d^2}{4}\right) = 28,575 \text{ lb}.$$

A *compressive* force of **28,575 pounds** is required on the bar to keep its length exactly 6 feet during installation.

coefficients, viscosity) that enable us to predict (and thus design for) the response of materials under load.

We next consider the situation where more than one material is loaded at a time, as well as changes in the stress-strain relationship over time.

VII. Problems

1. A 4.0-meter long, hollow cylindrical steel column will support a mass of 100,000 kilograms. The column has an outer diameter of 50 centimeters and an inner diameter (of the hollow portion) of 45 centimeters.

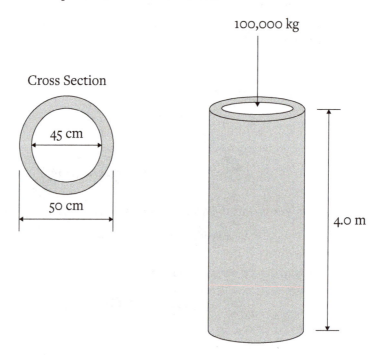

100,000 kg

Cross Section

45 cm

50 cm

4.0 m

   a. Based on this supported mass, what is the stress in the steel, in kPa?

   b. If the column compresses by 0.13 millimeters under this load, what is the axial strain?

2. A 1-inch by 1-inch square steel bar is 12 inches long. The bar is loaded in tension so that it experiences an axial strain of 0.5 percent.

   a. At this strain, what are the new dimensions and cross-sectional area of the bar if Poisson's ratio is 0.3?

   b. E = 29,000 ksi for the steel. Based on the original bar dimensions and 0.5 percent axial strain, what tensile stress exists in the bar assuming linear-elastic behavior?

   c. What tensile force is required to produce the stress calculated in part (b)?

3. A 48-inch-long aluminum bar (E = 10,000 ksi) consists of two, 24-inch long sections. Section AB is 4.0 inches in diameter, and section BC is 2.5 inches in diameter. The

bar is loaded in tension by a force, P = 25 kips. Determine the total elongation of the aluminum bar under this load.

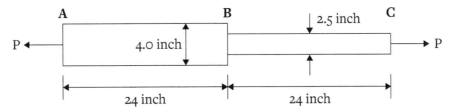

4. The following data were obtained from a tension test. The specimen's cross-section is circular (0.200 inches in diameter), and its original length was 3.000 inches. Assume the cross-section stays constant during the loading.

Load (lb)	Elongation (in)
0	0.0000
40	0.0010
84	0.0020
128	0.0030
168	0.0040
208	0.0050
248	0.0060
288	0.0070
320	0.0080
332	0.0090
344	0.0100
348	0.0120
350	0.0140
352	0.0180

a. Convert loads to stresses (psi) and elongation to strain. Make a table to show this result.

b. Plot the stress-strain diagram.

c. Determine the strain energy density and the strain energy from 0 percent to 0.2 percent strain.

d. Determine the yield strength for 0.1 percent offset.

e. Determine the ultimate strength of the material.

f. Determine the initial tangent modulus of elasticity.

g. Determine the secant modulus for an axial strain of 0.4 percent.

5. An A36 W-beam steel guardrail has an allowable strength of 165 MPa and E = 200 GPa. The W-beam section has a cross-sectional area of 14.35 centimeters² and a length of 4.0 meters. If a vehicle collides with the end of the guardrail such that it is loaded axially along its length, what is the maximum kinetic energy the vehicle can impart to the guardrail before it exceeds its allowable strength?

6. A building column applies a lateral load to a 5-centimeter thick aluminum baseplate that is 75 centimeters by 75 centimeters in plan. The bottom of the baseplate is affixed to the foundation and will not undergo displacement. If the lateral load imparted on the baseplate from the column is 10,000 kN and G = 27 GPa, what is the shear displacement that the baseplate will experience, in millimeters?

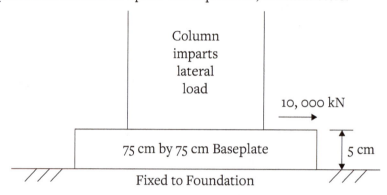

7. A bearing pad for a steel brace is made of two steel plates "sandwiching" an elastomeric pad (see figure). The pad is subjected to a shear force, V, during a static loading test. The pad has dimensions a = 12 inches and b = 13 inches, and the elastomer has a thickness, t = 1.50 inches. When the shear force, V, equals 50 pounds, the top plate is found to have displaced laterally 0.24 inches with respect to the bottom plate. Determine the shear modulus (G) of the elastomeric pad.

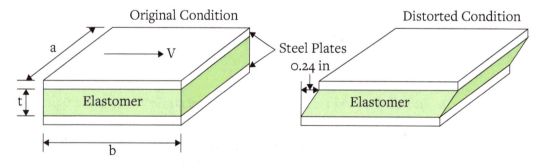

8. A 2-meter-long beam is part of a boat yard's marine travel lift for launching and retrieving boats. The beam will be supported on the ends and will have a triangular cross-section with the dimensions shown. The lift is carrying a boat, resulting in an

applied load at the mid-span of the beam, causing a bending moment on the beam, M = 11.4 kN-m.

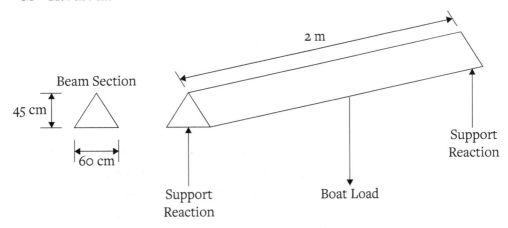

Beam Section

45 cm

60 cm

2 m

Support Reaction

Support Reaction

Boat Load

a. If the MOR for this beam is 3,100 kPa, what is the factor of safety (F.S.) for this situation?

$$FS = \frac{\text{Stress to Cause Failure}}{\text{Actual Applied Stress}}$$

b. Where on the beam's cross-section is the normal stress, $\alpha = 0$?

c. Where on the beam's cross-section is the normal stress a maximum? Compute this stress.

9. You are analyzing a 12-foot-long concrete beam with a rectangular cross-section with the width at 2.5 feet and depth at 4.4 feet. A maximum moment of 1.3×10^6 foot pounds is applied to the beam due to a bending load acting at the mid-span of the beam.

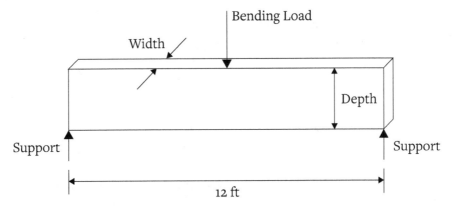

Bending Load

Width

Depth

Support

Support

12 ft

a. What is the bending load to induce this moment?

b. Measuring from the top of the beam, where on the beam's cross-section is the normal stress = 0, in feet?

c. What is the maximum tensile stress on the beam, in psi, and where on the beam's cross-section does it occur?

d. Determine the normal stress on the beam's cross-section 1.2 feet below the top of the beam.

10. The exhaust manifold on a car engine needs to be replaced. The manifold is held in place by one-fourth-inch diameter steel bolts that pass through holes in the manifold and thread into the cast iron engine block (i.e., the bolts are free to rotate within the holes in the manifold). The "free" length of the bolts (the length not screwed into the engine block) is 4.0 inches. Due to time and exposure to the elements during the operation of the car, the bolts have rusted in the engine block. To try to remove the bolts, a mechanic will apply a force on the end of a 1.0-foot-long wrench handle. The mechanic doesn't want to break the bolts during the process of trying to remove them, so he did some research and estimated the shear strength of the bolts is 75,000 psi and the shear modulus is $G = 11.5 \times 10^6$ psi.

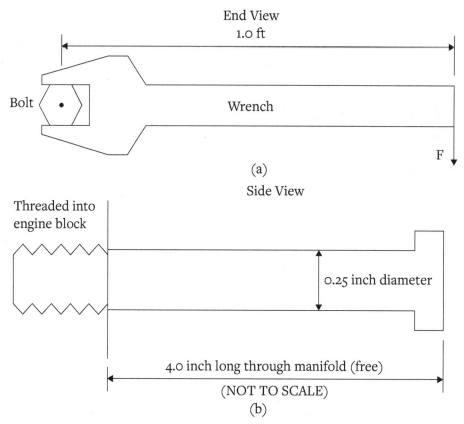

End View

1.0 ft

Bolt

Wrench

F

(a)

Side View

Threaded into engine block

0.25 inch diameter

4.0 inch long through manifold (free)

(NOT TO SCALE)

(b)

a. If the shear stress induced in the bolt from the wrench is not to exceed the shear strength (thus breaking the bolt), what is the maximum force the mechanic can apply at the end of the wrench handle?

b. If the bolts turn when the mechanic applies 15 pounds of force to the wrench handle, what is the maximum shear stress and shear strain in the bolt? What is the angle of twist that will be observed at the wrench, in degrees?

11. Copper pipe is used to plumb a baseboard-forced hot water heating system. When a 3.0-meter long section of baseboard heating was installed, the ambient temperature of all materials was 20°C. When the heat is running, hot water with a temperature of 85°C flows through the pipe.

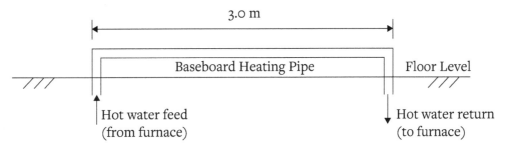

a. Based on this change in temperature, will the copper pipe lengthen or shorten?

b. If α_L = 17 × 10^{-6} (°C)$^{-1}$ for copper, how much will the 3.0-meter section of baseboard heating pipe change in length, in millimeters?

c. Should the holes in the floor for the feed and return pipes be drilled exactly the diameter of the pipe, or slightly oversized? Why?

Credit

Fig. 2.1: Source: https://commons.wikimedia.org/wiki/Isaac_Newton#/media/File:Sir_Isaac_Newton_(1643-1727).jpg.

3

Composite Models and Viscoelasticity

I. Composite Materials

The previous chapter considered the mechanical behavior (loading and deformation, stress and strain) of individual materials (i.e., components made of a single material type, such as steel or wood). When there is only one material type, only one set of material parameters (e.g., Young's modulus, E, or coefficient of thermal expansion, α) is necessary to solve the problem. However, there are multiple instances where engineers combine two (or more) materials in a component, resulting in a *composite material*.

Composite materials (also shortened to just composites) are materials made from two or more constituent materials with significantly different physical, chemical, or mechanical properties that, when combined, produce a material with characteristics different from the individual components. The most popular example in civil engineering is reinforced concrete members. Concrete is weak in tension (which will be discussed further in **chapter 6**). To increase the strength, a concrete member (such as a beam or column) is reinforced with steel rebar, which helps to take some of the tensile load. The number and size of the rebars are precisely designed based on the requirements of the concrete section. As engineers, we want to be able to predict the resulting mechanical properties of the section if we know the behavior of the individual materials in the composite system.

Typically, the individual components remain separate and distinct within the finished structure. That is, if you cut a concrete section in half, you can see the individual rebars in the cross-section. You could also inspect a sheet of plywood—individual layers (called plys) are easily distinguished along the edges. This distinct separation allows for prediction of mechanical properties of multi-material systems based on common composite rules (also called "rule of mixtures").

To understand the basics, we can first look at the behavior of a simpler "composite" system: springs.

Spring Models

In the previous chapter, springs were a useful tool to help explain the simple mechanical relationship between load and displacement (i.e., $F = k\delta$) and stress and strain (i.e., $\sigma = E\varepsilon$) when a simple spring was substituted for a material. Likewise, multiple springs can easily be used as a simple model for a multi-material—or composite—system. The key is recognizing that basic rules of force equilibrium and displacement compatibility must hold for the spring configuration.

We start by looking at a simple two-spring system in either parallel (linked alongside) or serial (linked end-to-end) configuration.

Springs in Parallel

We first consider two springs in parallel (linked alongside each other), with spring stiffness k_1 and k_2, subject to stretching (**figure 3.1**):

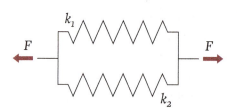

FIGURE 3.1   A system of two springs in parallel.

Hooke's law governs each spring, where $F_i = k_i \delta_i$. We know that, because of the boundary conditions, the deformation or displacement of each spring has to be equal/constant, or

$$\delta = \delta_1 = \delta_2.$$

This is the compatibility condition. If this condition is not met, then the springs would not meet at the ends.

Next, we consider the equilibrium condition that the forces in the system must balance. If, for example, we "cut" the spring system down the middle and consider the left side, then the force, F, pulling to the left must equal the internal forces in spring 1 and spring 2 pulling to the right (and the opposite for the right side of the system); the forces are **additive**. Thus, by equilibrium conditions, we can write

$$F = F_1 + F_2.$$

Moreover, in order to stretch to the displacement, δ, we need a force in each spring that can stretch spring 1 by δ_1 and spring 2 by δ_2. Substituting Hooke's law for each spring force,

$$F = k_1 \delta_1 + k_2 \delta_2,$$

for which we can substitute the known constant displacement, δ, and rearrange

$$F = (k_1 + k_2)\delta,$$

which can be reduced to

$$F = k_{eff}\delta,$$

where

$$k_{eff} = k_1 + k_2.$$

Thus, the effective spring stiffness of springs in parallel is simply the summation of the individual spring stiffness in the system. This holds in general, regardless of the number of springs, and can be written as

$$k_{eff,parallel} = \sum_{i=1}^{n} k_i,$$

where n is the number of springs in parallel.

Springs in Series

We next consider two springs in series (i.e., linked end to end), with spring stiffness k_1 and k_2 subject to stretching (**figure 3.2**):

k_1 k_2

F F

FIGURE 3.2   A system of two springs in series.

Again, each spring is governed by Hooke's law. This time, we know that because of the equilibrium condition, the load or force in each spring has to be equal, or

$$F = F_1 = F_2.$$

This can easily be shown if we cut the system anywhere and balance the forces.

We next consider the compatibility condition. Stretching the total displacement, δ, includes both the deformation of spring 1 by δ_1 and the deformation of spring 2 by δ_2; this time the displacements are **additive**. Thus, by compatibility condition, we can write

$$\delta = \delta_1 + \delta_2.$$

Again, substituting Hooke's law for each spring displacement,

$$\delta = \frac{F_1}{k_1} + \frac{F_2}{k_2},$$

for which we can substitute the known constant force, F, and rearrange as

$$\delta = \left(\frac{1}{k_1} + \frac{1}{k_2} \right) F.$$

And then we can rearrange a little more to resemble Hooke's law:

$$F = \left(\frac{1}{k_1} + \frac{1}{k_2}\right)^{-1} (\delta) = \left(\frac{k_1 k_2}{k_1 + k_2}\right)\delta,$$

which again can be reduced to

$$F = k_{eff}\delta,$$

where

$$k_{eff} = \left(\frac{1}{k_1} + \frac{1}{k_2}\right)^{-1}.$$

Here, the effective spring stiffness of springs in series is more complex—the inverse of the summation of the inverse of the spring stiffness. While complicated to say, this can be written simply as

$$\frac{1}{k_{eff,series}} = \sum_{i=1}^{n} \frac{1}{k_i},$$

where n is the number of springs in series.

Spring Compliance

While the concept behind springs in series is as simple as the parallel case, the math gets a little more complicated because of all the inverse stiffness terms. This can be avoided by introducing the compliance of a spring, c, which is defined as the inverse of the stiffness:

$$\text{compliance} = c = \frac{1}{k} = \frac{1}{\text{stiffness}}.$$

If you think of stiffness as the necessary force required to deform a unit displacement (e.g., Newton per meter), then the compliance can be thought of as the reverse—the resulting displacement owing to a unit force (meters per Newton). This leads to the compliance version of Hooke's law, where

$$\delta = cF,$$

which can easily be reformulated as $F = k\delta$. Using compliances and the alternative form of Hooke's law noted earlier, it can be easily shown that

$$c_{eff,series} = \sum_{i=1}^{n} c_i.$$

and

$$\frac{1}{c_{eff,parallel}} = \sum_{i=1}^{n} \frac{1}{c_i}.$$

That is, the effective compliances for springs in series and parallel follow the opposite formulation than the effective stiffness.

A summary of parallel and serial spring effective stiffness equations is given in **table 3.1**.

TABLE 3.1 EFFECTIVE SPRING STIFFNESS

System	Number of springs	Stiffness formulation	Compliance formulation
Parallel	2	$k_{eff,parallel} = k_1 + k_2$	$c_{eff,parallel} = \dfrac{c_1 c_2}{c_1 + c_2}$
	n	$k_{eff,parallel} = \sum_{i=1}^{n} k_i$	$\dfrac{1}{c_{eff,parallel}} = \sum_{i=1}^{n} \dfrac{1}{c_i}$
Series	2	$k_{eff,series} = \dfrac{k_1 k_2}{k_1 + k_2}$	$c_{eff,series} = c_1 + c_2$
	n	$\dfrac{1}{k_{eff,series}} = \sum_{i=1}^{n} \dfrac{1}{k_i}$	$c_{eff,series} = \sum_{i=1}^{n} c_i$

Observations of Spring Models

We now have some simple relations to determine the effective stiffness of springs in parallel and series. What can we learn from the relations in general?

Consider again two springs in parallel. What happens if one spring is much stiffer than the other? That is, what if

$$k_1 \gg k_2.$$

Can we make any claims about k_{eff}? For arguments sake, let $k_1 = 1,000$ N/m and $k_2 = 1$ N/m. Then, by the known relation for *parallel* springs,

$$k_{eff,parallel} = k_1 + k_2 = 1,000 + 1 = 1,001 \text{ N/m}.$$

We see that k_{eff} is slightly larger than k_1. In fact, we can clearly see that, as $k_2 \to 0$, then $k_{eff} \to k_1$. In general, we can make the conclusion that for springs in parallel, *the effective total stiffness is always greater than (or equal to) the largest spring stiffness in the system*. In other words, the stiffest spring (in our simple example, $k_1 = 1,000$ N/m) sets the *minimum total stiffness* of the parallel system. Note that this holds regardless of the number of springs, n.

Now, consider two springs in series. Again, what happens if one spring is much stiffer than the other? Let us again set $k_1 = 1,000$ N/m and $k_2 = 1$ N/m. Then, by the known relation for *serial* springs,

$$k_{eff} = \left(\frac{1}{k_1} + \frac{1}{k_2}\right)^{-1} = \left(\frac{1}{1,000} + \frac{1}{1}\right)^{-1} = (1.001)^{-1} \cong 0.999 \text{N}/\text{m}.$$

In contrast to the parallel case, we see that k_{eff} is slightly smaller than k_2. Here, we can clearly see that, as $k_1 \to \infty$, then $1/k_1 \to 0$ and $k_{eff} \to k_2$. That is, the trend is opposite to the parallel case. In general, we can make the conclusion that for springs in series, *the effective total stiffness is always less than (or equal to) the smallest spring stiffness in the system.* In other words, the most compliant/weakest spring (in our simple example, $k_2 = 1$ N/m) sets the *maximum total stiffness* of the serial system. Note that, again, this holds regardless of the number of springs, n.

What happens in a parallel system if one of the spring stiffnesses approaches infinity? Is this a physical case? When could it happen? Likewise, what happens in a serial system if one of the springs approach zero stiffness? Is this a physical case? When could it happen?

Combining Serial and Parallel Models

A system of more than two springs is not limited to purely parallel or purely serial arrangements. Indeed, one could combine the springs in many configurations. However, assuming a linear or horizontal arrangement, the system can always be broken down into smaller subsystems of parallel or serial springs. The key is to work out the effective stiffness of each subsystem and reduce the total system sections at a time.

For example, consider the system of five springs as shown in **figure 3.3**. Springs 1 and 2 are in series, which are both in parallel with spring 3. Springs 4 and 5 are in series, and springs 1, 2, and 3 (combined) are in series with springs 4 and 5.

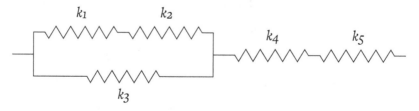

FIGURE 3.3   A system of five springs.

To replace with a single total effective stiffness:

Step 1: Replace springs 1 and 2 with spring 6:

$$k_6 = \left(\frac{1}{k_1} + \frac{1}{k_2} \right)^{-1} = \frac{k_1 k_2}{k_1 + k_2}.$$

Step 2: Replace springs 4 and 5 with spring 7:

$$k_7 = \left(\frac{1}{k_4} + \frac{1}{k_5} \right)^{-1} = \frac{k_4 k_5}{k_4 + k_5}.$$

Step 3: Replace springs 3 and 6 with spring 8:

$$k_8 = k_6 + k_3 = \frac{k_1 k_2}{k_1 + k_2} + k_3 = \frac{k_1 k_2 + k_1 k_3 + k_2 k_3}{k_1 + k_2}.$$

Step 4: Replace springs 7 and 8 with the total spring stiffness:

$$k_{eff} = \left(\frac{1}{k_7} + \frac{1}{k_8} \right)^{-1} = \left(\frac{k_4 + k_5}{k_4 k_5} + \frac{k_1 + k_2}{k_1 k_2 + k_1 k_3 + k_2 k_3} \right)^{-1}$$

or

$$k_{eff} = \frac{k_1 k_2 k_4 k_5 + k_1 k_3 k_4 k_5 + k_2 k_3 k_4 k_5}{k_1 k_2 k_4 + k_1 k_3 k_4 + k_2 k_3 k_4 + k_1 k_2 k_5 + k_1 k_3 k_5 + k_2 k_3 k_5 + k_1 k_4 k_5 + k_2 k_4 k_5}.$$

While the algebra gets a bit tangled, the concept is the same. As an exercise, you can make up numerical values for the initial five springs and work out the validity of the earlier relation. With some effort, even the most complex spring systems can be solved using a systematic piecewise approach. Note that, as long as the subsystems are valid, it doesn't matter what order you group and reduce the springs; the total effective stiffness will be the same.

Now we know a little about multiple spring systems. However, that is only the first step in understanding composite material systems. Just like the previous chapter, we must now convert the spring values (force and displacement) to more general material values (stress and strain).

Composite Material Rules of Mixture

We can easily transfer the "spring perspective" to composite material systems by first assuming that materials are combined in either a parallel or serial fashion. From this, both equilibrium (force balance) and compatibility (deformation) conditions are still applicable.

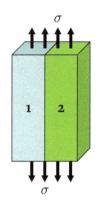

FIGURE 3.4 Materials in a parallel system.

Note that, for these simple cases to hold, we assume the materials are isotropic, linear elastic, and perfectly bonded to one another at the interface. If these conditions are not met, more complex composite material rules must be used.

Parallel Materials

First, we consider two materials in a parallel system—that is, a load is applied in which stress is *transverse* to the materials at their interface (see **figure 3.4**). Each material is linear elastic, with a Young's modulus E, and governed by Hooke's law, where $\sigma_i = E_i \varepsilon_i$.

Similar to parallel springs, we know that, because of the boundary conditions, the deformation or displacement of each material has to be equal/constant, or

$$\delta = \delta_1 = \delta_2.$$

If this condition is not met, then the material would not meet at the ends. However, now we wish to convert the displacements to strains. We know that, by definition, $\varepsilon = \delta/L$; therefore, $\delta = \varepsilon L$, and substituting for the earlier equation,

$$\varepsilon L = \varepsilon_1 L_1 = \varepsilon_2 L_2.$$

In this case, the lengths in the direction of the load are identical for each material such that $L = L_1 = L_2$, which means

$$\varepsilon = \varepsilon_1 = \varepsilon_2.$$

This is the strain compatibility condition. Note the similarity to the springs in parallel condition. For materials in parallel, the *strain is constant* (otherwise the material components would separate wherever they are connected).

Next, we consider the equilibrium condition that the *forces* in the system must balance. We note here that you cannot balance stresses! Forces are subject to equilibrium, and as such, like the spring model, the forces are **additive**. The stresses are not additive! Thus, by equilibrium conditions, we write

$$F = F_1 + F_2.$$

We can, however, convert the forces to stress by considering the area. We know that $\sigma = F/A$ and thus $F = \sigma A$. Substituting for forces in terms of stresses and areas into the previous equation, we have

$$\sigma A = \sigma_1 A_1 + \sigma_2 A_2.$$

Furthermore, we can relate stress and strain via Young's modulus, E, where $\sigma = E\varepsilon$ such that

$$E\varepsilon A = E_1 \varepsilon_1 A_1 + E_2 \varepsilon_2 A_2.$$

Since the strains are constant (compatibility condition), this can be reduced to

$$EA = E_1 A_1 + E_2 A_2.$$

Since we are looking for the effective modulus, we can rewrite the earlier equation as

$$E = E_1 \left(\frac{A_1}{A} \right) + E_2 \left(\frac{A_2}{A} \right).$$

Thus, the effective modulus of the composite system with materials in parallel is reduced to the summation of the individual material's modulus, weighted by the proportional area in the system's cross-section. Note that

$$A = A_1 + A_2$$

such that

$$\frac{A_1}{A} + \frac{A_2}{A} = 1;$$

that is, all the area weights must sum to one.

If there are more than two materials in parallel, this relationship can be written in general for n materials as

$$E_{eff,parallel} = \frac{1}{A} \sum_{i=1}^{n} E_i A_i.$$

We next consider materials in series.

Serial Materials

For materials in series, the materials are aligned end to end such that the applied load is primarily along the major axis of each material and is *normal* to the material interface (see **figure 3.5**). Again, each material is linear elastic, with a Young's modulus E, and governed by Hooke's law.

This time, we know that because of the equilibrium condition, the load or force in each material has to be equal, or

$$F = F_1 = F_2.$$

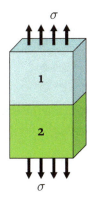

FIGURE 3.5 Materials in a serial system.

This can again easily be shown if we cut the system anywhere and balance the forces. Again, we wish to convert the forces to stresses:

$$\sigma A = \sigma_1 A_1 = \sigma_2 A_2.$$

This time, the cross-sectional area normal to the applied load is constant, or $A = A_1 = A_2$ such that

$$\sigma = \sigma_1 = \sigma_2.$$

In series, the stress in each material component is constant (assuming the same cross-sectional area). This is the stress equilibrium condition. Note the similarity to springs in series.

We next consider the compatibility condition. Upon stretching, the total displacement, δ, includes both the deformation of material 1 and the deformation of material 2. Again, the displacements are **additive**. We cannot add strains of two different materials! By compatibility condition, we can write

$$\delta = \delta_1 + \delta_2.$$

Using the known strain relation, we can substitute and get

$$\varepsilon L = \varepsilon_1 L_1 + \varepsilon_2 L_2.$$

Again, using the relationship between stress and strain, where $\sigma = E\varepsilon$, we can substitute Young's modulus in the earlier equation, arriving at the following:

$$\left(\frac{\sigma}{E}\right)L = \left(\frac{\sigma_1}{E_1}\right)L_1 + \left(\frac{\sigma_2}{E_2}\right)L_2.$$

Since the stress is constant (equilibrium condition), then the aforementioned reduces to

$$\left(\frac{1}{E}\right)L = \left(\frac{1}{E_1}\right)L_1 + \left(\frac{1}{E_2}\right)L_2.$$

Isolating the effective modulus results in

$$E = \left(\left(\frac{1}{E_1}\right)\left(\frac{L_1}{L}\right) + \left(\frac{1}{E_2}\right)\left(\frac{L_2}{L}\right)\right)^{-1}.$$

Thus, the effective modulus of the composite system with materials in series is reduced to the inverse of the summation of the inverse of each individual material's modulus, weighted by the proportional area in the system's length. Note that for series

$$L = L_1 + L_2,$$

$$\frac{L_1}{L} + \frac{L_2}{L} = 1;$$

that is, all the length weights must sum to one.

If there are more than two materials in series, this relationship can be written in general for n materials as

$$E_{eff,series} = \left(\frac{1}{L} \sum_{i=1}^{n} \frac{L_i}{E_i} \right)^{-1}.$$

Volume Fractions

We see that in parallel composite systems, the contribution to the modulus is weighted by proportional areas, or A_i / A, where the length in the direction of the load, L, is constant. If we multiply the ratio by L/L, then

$$\frac{A_i}{A} \left(\frac{L}{L} \right) = \frac{V_i}{V};$$

that is, we converted the proportional area to the proportional volume.

In serial composite systems, the contribution to the modulus is weighted by proportional length, or L_i / L, where the cross-sectional area normal to the direction of the load, A, is constant. If we multiply the ratio by A/A, then

$$\frac{L_i}{L} \left(\frac{A}{A} \right) = \frac{V_i}{V};$$

that is, we converted the proportional length to the proportional volume.

For example, a composite may consist of two alternating materials in series. The length of each "layer" in the composite can be combined and converted into a proportional volume of the entire composite. This relationship is shown in **figure 3.6**.

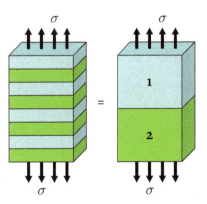

FIGURE 3.6 Volume fraction illustration for a serial composite; several layers of two alternating materials represented as proportional volumes of the composite.

This conversion is useful, as now we do not need to consider areas or lengths in our calculation of effective modulus—only volume ratio, or volume fraction, f, where

$$f_i = \frac{V_i}{V},$$

where

$$V = \sum_{i=1}^{n} V_i$$

such that

$$\sum_{i=1}^{n} f_i = 1.$$

With the concept of volume fraction, we can rewrite our composite modulus equations as

$$E_{eff,parallel} = \sum_{i=1}^{n} E_i f_i$$

and

$$E_{eff,series} = \left(\sum_{i=1}^{n} \frac{f_i}{E_i} \right)^{-1}.$$

A summary of parallel and serial effective composite modulus equations is given in **table 3.2.**

TABLE 3.2 EFFECTIVE COMPOSITE MODULUS

System	Number of Springs	Stiffness Formulation
Parallel	2	$E_{eff,parallel} = E_1 f_1 + E_2 f_2$
	n	$E_{eff,parallel} = \sum_{i=1}^{n} E_i f_i$
Series	2	$\dfrac{1}{E_{eff,series}} = \dfrac{f_1}{E_1} + \dfrac{f_2}{E_2}$
	n	$E_{eff,series} = \left(\sum_{i=1}^{n} \dfrac{f_i}{E_i} \right)^{-1}$

Theoretical Bounds

When we considered springs, we concluded two things: (1) for springs in parallel, the effective total stiffness is always greater than (or equal to) the largest spring stiffness in the system, and (2) for springs in series, the effective total stiffness is always less than (or equal to) the smallest spring stiffness in the system. Does this hold for composite materials?

Unfortunately, because of the variability in volume fraction, this does not hold. For example, consider two materials in parallel, consisting of 50 percent volume each, where one material has a modulus of 100 GPa and the other 50 GPa. The modulus of the composite system is 75 MPa, which lies between the two values (note that this cannot happen for springs!). As an exercise, you can determine the composite modulus if the materials were in a serial configuration.[1]

Can we make any general conclusions about these rules of mixture? Yes.

Consider a simple two-phase composite material consisting of material "A" and material "B." Let the volume fraction of material "A" be $f_A = f$, and, because $f_A + f_B = 1$, then $f_B = 1 - f$. First, consider the materials in parallel. The effective modulus is

$$E_{AB,parallel} = E_A f_A + E_B f_B = f E_A + (1-f) E_B.$$

"Massaging" this equation a little results in

$$E_{AB,parallel} = (E_A - E_B)f + E_B,$$

which is an expression for the effective modulus, $E_{AB,parallel}$ as a function of the volume fraction of "A," f, rearranged to a linear form (i.e., $y = mx + b$) where the slope (m) is simply "$E_A - E_B$" and the intercept (b) is "E_B." When $f = 1$, the system is 100 percent material "A" and, from the earlier relationship, $E_{AB,parallel} = E_A$. Likewise, when $f = 0$, the system is 100 percent material "B" and, from the aforementioned, $E_{AB,parallel} = E_B$. Any value of f from zero to one results in a linear interpolation between E_A and E_B. This seems logical for "mixing" two materials. We can plot this relationship (**figure 3.7**). Note that the "slope" of the relationship changes according to the magnitudes of E_A and E_B.

Now, let us consider the same material is a serial configuration. The effective modulus is

$$E_{AB,series} = \left(\frac{f_A}{E_A} + \frac{f_B}{E_B} \right)^{-1} = \left(\frac{f}{E_A} + \frac{1-f}{E_B} \right)^{-1}.$$

Manipulating this relationship, we find

$$E_{AB,series} = \left(\frac{(E_B - E_A)f + E_A}{E_A E_B} \right)^{-1} = \frac{E_A E_B}{(E_B - E_A)f + E_A}.$$

The aforementioned relation is not as easy to recognize. It is still an expression for the effective modulus, $E_{AB,series}$, as a function of the

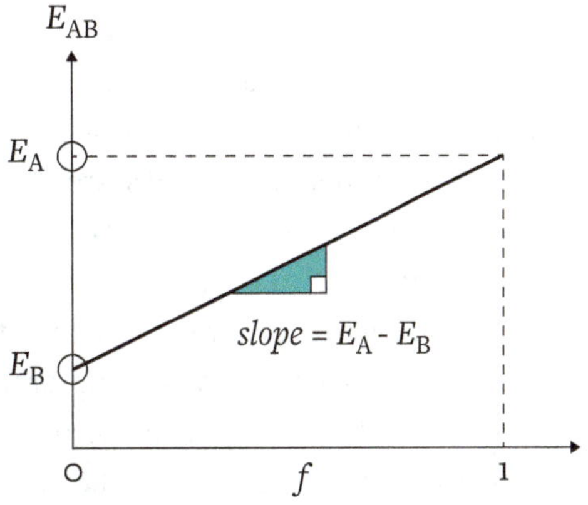

FIGURE 3.7   Relationship for the effective modulus of a composite material composed of materials A and B, assuming the materials are in parallel.

---

1   In series, the effective modulus would be 66.67 GPa.

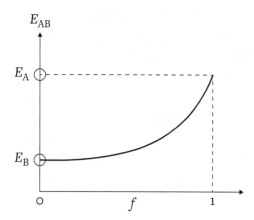

E_{AB}

E_A

E_B

O f 1

FIGURE 3.8 Relationship for the effective modulus of a composite material composed of materials A and B, assuming the materials are in series.

volume fraction of "A," f. However, rather than a simple linear relation, it is a type of inverse relationship with respect to f. For example, it is similar to $y = 1/x$, as the independent variable is in the denominator; there are just a bunch of other constants.

When $f = 1$, the system is 100 percent material "A" and, from the aforementioned relationship, $E_{AB,series} = E_A$ (check!). Likewise, when $f = 0$, the system is 100 percent material "B" and, from the aforementioned, $E_{AB,series} = E_B$ (check!). Values of f from zero to one result in some intermediate modulus values, but getting from E_B to E_A is not as clear as the prior case. Just for fun, what does the denominator look like? The denominator is a linear relation, like the expression for parallel systems, but with the "A" and "B" contributions swapped.

To get a clearer picture of what is happening, we can plot the relationship (see **figure 3.8**).

We see that—for every value of f—the modulus for materials in parallel exceeds the modulus for the same materials with the same volumetric proportions in series. This, in fact, is the case for any number of materials ($n = 2, 3, 4, ...$) and any value of moduli. Thus, the parallel case defines the **upper bound** for any combination of materials, while the serial case defines the **lower bound** for any combination of materials. We note that in some texts, the parallel case is sometimes referred to as the **Voigt model**, whereas a serial model is referred to as the **Reuss model**, and the two together define the **Voigt-Reuss bounds[2] for composites**.

Defining the theoretical maximum and minimum cases is extremely helpful for engineers in cases where (a) approximations can be made for more complicated geometries and (b) exact answers are not possible because of a lack of information.

We have considered material systems in both ideal parallel configurations and ideal serial configurations. But how can we handle a nonideal case? For example, consider an embedded circular shape, such as depicted in **figure 3.9(a)**. If the load is in the x-direction, are the materials in parallel or series? In effect, they are both parallel and serial at the same time! Depending on the geometry, solving for the composite modulus analytically is a complicated procedure involving integration across the system.

We can easily assume first that the materials are in parallel and determine the upper bound modulus. Next, we assume that the materials are in series and determine the lower-bound

2 These relations also define a theoretical upper and lower bound on other material properties, such as mass density, ultimate tensile strength, thermal conductivity, and electrical conductivity.

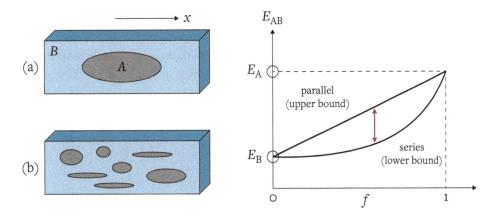

FIGURE 3.9   Examples of upper and lower bounds.

modulus. The actual modulus is somewhere in between. Depending on the situation and required accuracy (i.e., necessary engineering "exactitude"), we can either use an average as an approximation or use the worst-case scenario for a calculation.

If a more accurate modulus is needed, numerical methods, such as finite elements, are usually employed. Even then, the upper and lower bounds can be calculated as a "sanity check" for the computational solution. Remember, computers are not correct 100 percent of the time.

The same consideration can be applied to a system of unknown consideration, such as depicted in **figure 3.9(b)**. Instead of a known geometric shape, some reinforcing material may be randomly distributed across a composite system. This time, not only is the system both parallel and serial, but also it can neither be modeled analytically nor numerically. For such cases, the theoretical bounds are the only method we have to approximate the modulus.

Order Matters

We stated that in a system of springs, one can systematically reduce the stiffness to a single spring constant by breaking down the system into sections of parallel and serial configurations. Moreover, the order in which you break down the system is arbitrary; you will resolve the same effective stiffness.

Is this the case for composite materials? Consider the system depicted in **figure 3.10**. It consists of a single,

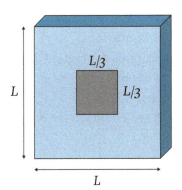

FIGURE 3.10   Composite material system used to demonstrate the difference in effective modulus based on the method used to reduce the system.

square inclusion at the center of a plate, which is broken up into thirds. Thus, the inclusion material represents one-ninth of the volume, while the bulk material represents eight-ninths of the volume. In addition, the inclusion material has a modulus of ten. The bulk material has a modulus of one (units do not matter in this case). Assume there is a load in the horizontal direction. Let us reduce the modulus via four different approaches and compare.

Case 1: Upper bound, assume parallel

$$E_{parallel} = E_{bulk}f_{bulk} + E_{inclusion}f_{inclusion} = 1\left(\frac{8}{9}\right) + 10\left(\frac{1}{9}\right) = \frac{18}{9} = 2$$

Case 2: Lower bound, assume series

$$E_{series} = \left(\frac{f_{bulk}}{E_{bulk}} + \frac{f_{inclusion}}{E_{inclusion}}\right)^{-1} = \left(\frac{\frac{8}{9}}{1} + \frac{\frac{1}{9}}{10}\right)^{-1} = \left(\frac{81}{90}\right)^{-1} = \frac{90}{81} \cong 1.11$$

Comparing these cases, we have an upper bound of 2 and a lower bound of 1.11, which is acceptable for initial calculations. However, we can do better if we break down the system a little further.

Case 3: Break the system down to three parallel springs consisting of serial configurations

Using this method, we first have two types of serial springs. The first (top and bottom) are the bulk material only, such that the effective modulus is simply one (no change) (see **figure 3.11**).

The second has the effect of the inclusion in series, such that

$$E_{middle} = \left(\frac{f_{bulk}}{E_{bulk}} + \frac{f_{inclusion}}{E_{inclusion}}\right)^{-1} = \left(\frac{\left(\frac{2}{3}\right)}{1} + \frac{\left(\frac{1}{3}\right)}{10}\right)^{-1}$$

$$= \left(\frac{21}{30}\right)^{-1} = \frac{30}{21} \cong 1.43.$$

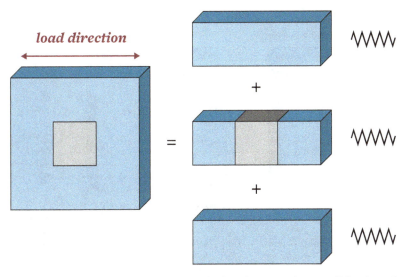

FIGURE 3.11   Composite material system broken down into three parallel springs of serial configurations.

Note that, since we are only considering the middle third of the whole system, the volume fractions have been modified to one-third and two-thirds for the inclusion and bulk, respectively. Now, this spring system is parallel with the other two, such that

$$E_{parallel} = E_{bulk}f_{bulk} + E_{middle}f_{middle} = 1\left(\frac{2}{3}\right) + \left(\frac{30}{21}\right)\left(\frac{1}{3}\right) = \frac{72}{63} \cong 1.14.$$

Now we have another approximation of the modulus of 1.14.

Case 4: Break the system down to three serial springs consisting of parallel configurations

This method is essentially the reverse of case 3 (see **figure 3.12**).

We first have two types of parallel springs. The first (left and right) are the bulk material, with a modulus of one (again, no change). The second has the effect of the inclusion in parallel, such that

$$E_{middle} = E_{bulk}f_{bulk} + E_{inclusion}f_{inclusion} = 1\left(\frac{2}{3}\right) + 10\left(\frac{1}{3}\right) = \frac{12}{3} = 4.$$

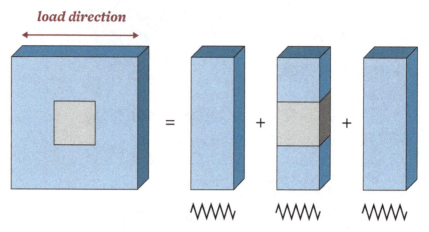

FIGURE 3.12   Composite material broken down into three serial springs of parallel configurations.

Again, the volume fractions have been modified to one-third and two-thirds for the inclusion and bulk, respectively. Now, this spring system is in series with the other two, such that

$$E_{series} = \left(\frac{f_{bulk}}{E_{bulk}} + \frac{f_{middle}}{E_{middle}} \right)^{-1} = \left(\frac{\left(\frac{2}{3}\right)}{1} + \frac{\left(\frac{1}{3}\right)}{4} \right)^{-1}$$

$$= \left(\frac{9}{12} \right)^{-1} = \frac{12}{9} \cong 1.33.$$

This is different from our previous answer! We can conclude two things from these results:

1. The calculated composite modulus determined by breaking down a composite system into smaller sections of parallel and serial models is dependent on the order in which the system is reduced.

2. You can reduce the system to refine the upper and lower bounds of the modulus. By assuming serial then parallel, we approximated a modulus of 1.14; while assuming parallel then serial, we approximated a modulus of 1.33. The true modulus of the system will lie somewhere between these values, which is a much narrower range than the initial approximation of $1.11 < E < 2$, as determined by case 1 and case 2.

Note that it is not always clear which approach will result in the upper and lower bounds (dependent on the geometry of the system as well as the relative moduli).

Dealing with Voids

Many material systems have pores or voids within them. These voids can (sometimes) be considered a distinct material phase, with a stiffness approaching zero. See **figure 3.13**.

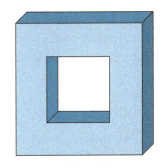

FIGURE 3.13   Material with void in the middle for determining effective modulus.

The modulus can be approximated using the parallel model, accounting for the drop-in material density owing to voids/porosity, where

$$E_{\text{material with voids}} = (1 - f_{\text{voids}})E_0,$$

where f_{voids} is the volume fraction of voids and E_0 the modulus of the bulk solid material. Note that this can also be approximated using density, ρ, as a variable, where

$$\rho_{\text{material with voids}} = (1 - f_{\text{voids}})\rho_0,$$

where ρ_0 is the density of the bulk solid material. As an exercise, try to derive the earlier relation. Substituting to the modulus relation,

$$E_{\text{material with voids}} = \left(\frac{\rho_{\text{material with voids}}}{\rho_0}\right)E_0.$$

Note that the serial model cannot be applied, because, similar to a spring with zero stiffness, the composite modulus approaches zero as one of the material moduli approaches zero. Thus, this approximation of the modulus of a porous material is the upper bound—it is the maximum stiffness attainable if the pores/voids were geometrically aligned to maximize stress transfer in the remaining material. Unfortunately, this is not usually the case, and many modifications have been made to approximate porous material stiffness based on additional parameters, such as average pore size and shape. This is beyond the scope of the current discussion.

With the simple rules of mixtures, we can approximate the modulus of many composite systems, derived from simple spring models. However, at the onset, we limited such materials to linear-elastic behaviors. Introduction of more complex behaviors requires the introduction of more complex elements to add to our spring representations. We will demonstrate such additional complexity via a brief discussion of the time dependence of materials, also called **viscoelasticity**.

II. Time-Dependent Response and Viscoelasticity

Thus far, all of the mechanical behaviors and responses we have discussed have been considered time independent. That is, an applied stress is instantaneously applied, and the deformation (be it normal or shear strain, bending, or twist) also occurs instantaneously. Likewise, when loads are removed, everything returns to the stress-free state immediately (notwithstanding elastic or plastic strains).

Many engineering materials (such as asphalt) will have both a stress- and time-dependent response. Even concrete, over long time spans, demonstrates some time dependence in either the stress or deformation. This time-dependent response is sometimes called the **viscous response** of a material. The viscosity of a material is a measure of its resistance to gradual deformation by shear stress or tensile stress. For liquids and fluids, it corresponds to the informal notion of "thickness." For example, honey has a higher viscosity than water.

The subject of how matter flows is a field called **rheology**. Rheology is primarily concerned with materials in the liquid state, but also as "soft solids" or solids under conditions in which they respond with plastic flow rather than deforming elastically in response to an applied force.

The Main Difference between Elastic and Viscous Behavior

We consider a simple thought experiment where we load two materials: (1) an elastic solid and (2) a viscous liquid with a constant stress over a time interval (**figure 3.14**):

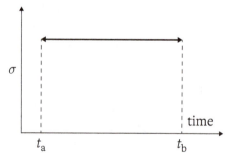

FIGURE 3.14   Load history for our thought experiment. At time t_a, a constant stress is applied (the magnitude does not matter). At time t_b, the load is removed.

The elastic solid behaves as we would expect (**figure 3.15**):

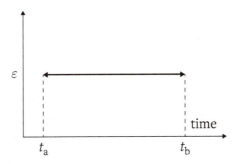

FIGURE 3.15 Strain history for our thought experiment, elastic solid. At time t_a, a constant strain is incurred instantaneously (the magnitude can be calculated via Hooke's law, where $\varepsilon = \sigma / E$). At time t_b, the load is removed, and the strain immediately drops to zero.

For the time in which the constant stress is applied (from t_a to t_b), the strain is also constant, where

$$\varepsilon = \frac{\sigma}{E}.$$

There is no dependence on time. The viscous liquid behaves differently (**figure 3.16**):

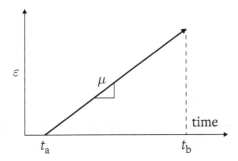

FIGURE 3.16 Strain history for our thought experiment, viscous liquid.

At time t_a, there is no strain. However, the strain increases constantly with time. The rate of strain increase, $\dot{\varepsilon}$, is related to the viscosity, μ, of the material. For the time in which the constant stress is applied (from t_a to t_b), the strain is constantly increasing, where

$$\dot{\varepsilon} = \frac{d\varepsilon}{dt} = \frac{\sigma}{\mu}.$$

We see a clear dependence on time (the "*dt*" term). In fact, this represents a simple differential equation. The key point here is that strain is not constant.

Simple Rheological Models for Solids

Viscoelasticity is the property of materials that exhibit both viscous and elastic characteristics when undergoing deformation. Viscous materials, like honey, resist shear flow and strain linearly with time when a stress is applied. Elastic materials strain when stretched and quickly return to their original state once the stress is removed. Viscoelastic materials have elements of both of these properties and, as such, exhibit time-dependent strain.

If we wish to analyze viscoelastic materials, we can begin by separating the behaviors. We can derive simple mechanical models to show different types of behavior and then use time in combinations to represent more complex mechanical behavior. **Table 3.3** gives the most common mechanical models.

The most commonly used elements for describing solids that are both elastic *and* viscous are the spring (elastic) and dashpot (viscous) models, and combinations thereof.

We are familiar with spring behavior, as we have been discussing it throughout this chapter.

The dashpot behaves like the viscous liquid in our thought experiment: it takes a while until there is any strain when it encounters load. This results in two limiting behaviors of the dashpot:

1. At time = zero, a dashpot is rigid. Regardless of the applied stress, there is no strain, and thus the initial "stiffness" of the dashpot is considered infinite.
2. As time approaches infinity, the dashpot opens or relaxes. It no longer carries any stress. Regardless of strain, there is no stress, and thus the final "stiffness" of the dashpot is considered zero.

TABLE 3.3 MECHANICAL MODEL ELEMENTS

Name	Element Schematic	Equation	Description
Hookean (spring)		$\sigma = E\varepsilon$	Linear stress-strain relation (e.g., Hooke's law), spring constant of Young's modulus, E.
Newtonian (dashpot)		$\sigma = \mu\dot{\varepsilon}$	Linear relation between stress and strain rate through constant viscosity, μ.
St. Venant (sliding block)		$\sigma < \sigma_{yield}, \varepsilon = 0$ $\sigma = \sigma_{yield}, \varepsilon \to \infty$	No deformation until yield stress is reached and then sliding at constant stress; cannot exceed σ_{yield}.

The simplest combination of the spring and dashpot models is either in series or parallel, and both are used so commonly that they have special designations:

1. Spring + Dashpot in series = Maxwell model
2. Spring + Dashpot in parallel = Kelvin/Voigt model

Maxwell Model

The Maxwell model is defined by a single spring element and a single dashpot element in series (**figure 3.17**). It is most applicable for materials that have some initial stiffness but relax to zero resistance over time.

FIGURE 3.17 Schematic of a Maxwell model consisting of a single spring element and a single dashpot element in *series*.

The element equations can be used to derive the governing equation using what we know about composite models. Since the system is in series, the displacements are additive, and the strains can be added since the material length is constant, such that

$$\varepsilon = \varepsilon_{spring} + \varepsilon_{dashpot}.$$

Since the dashpot requires the strain rate, we take a derivative with respect to time:

$$\dot{\varepsilon} = \frac{d}{dt}\varepsilon_{spring} + \dot{\varepsilon}_{dashpot}.$$

Now we can substitute the rules for our elements, $\sigma_{spring} = E\varepsilon_{spring}$ and $\sigma_{dashpot} = \mu\dot{\varepsilon}_{dashpot}$, solved for strain:

$$\dot{\varepsilon} = \frac{d}{dt}\left(\frac{\sigma_{spring}}{E}\right) + \frac{\sigma_{dashpot}}{\mu}.$$

Also, since the system is in series, the stresses for the elements are equal, such that

$$\sigma = \sigma_{spring} = \sigma_{dashpot}.$$

Since we don't have to differentiate between the stress in the spring or dashpot, we can simplify our equation:

$$\dot{\varepsilon} = \frac{d}{dt}\left(\frac{\sigma}{E}\right) + \frac{\sigma}{\mu}.$$

And, finally, we take the derivative of the stress (note this assumes the modulus is independent of time, which is fine for most engineering materials) and attain the governing equation for the Maxwell model:

$$\dot{\varepsilon} = \frac{\dot{\sigma}}{E} + \frac{\sigma}{\mu}.$$

There are two distinct loading situations to account for:

1. **Creep:** Apply a constant load/stress at t = 0 and leave it. The load will cause the dashpot to extend over time, also called creep.
2. **Relaxation:** Apply a constant strain at t = 0 and leave it. The internal stress in spring will relax as the dashpot undergoes strain.

For the Maxwell model, creep is characterized by the following (**figure 3.18**):

a. An instantaneous elastic strain (because of the spring element)
b. Linearly increasing strain with time (because of the dashpot)

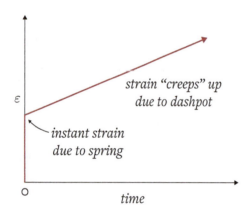

strain "creeps" up due to dashpot

instant strain due to spring

FIGURE 3.18   Maxwell model creep.

We can consider our governing equation,

$$\dot{\varepsilon} = \frac{\dot{\sigma}}{E} + \frac{\sigma}{\mu},$$

with the conditions that $\dot{\sigma} = 0$. Then our differential equation reduces to

$$\dot{\varepsilon} = \frac{\sigma}{\mu},$$

which is a relatively easy equation to solve. Integrating both sides with respect to time,

$$\varepsilon(t) = \frac{\sigma}{\mu}t + C.$$

We also know that at t = 0, the strain is equal to that of the spring only, or $\varepsilon(0) = \varepsilon_0 = \sigma/E = C$, and we arrive at the strain creep equation for the Maxwell model:

$$\varepsilon(t) = \left(\frac{\sigma}{\mu}\right)t + \left(\frac{\sigma}{E}\right).$$

1. For the Maxwell model, relaxation is characterized by the following (**figure 3.19**):

    a. An instantaneous stress (because of the spring element)
    b. Exponential decreasing stress (because of the dashpot).

We again consider the governing equation

$$\dot{\varepsilon} = \frac{\dot{\sigma}}{E} + \frac{\sigma}{\mu}.$$

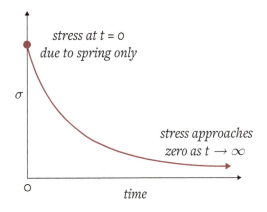

FIGURE 3.19   Maxwell model relaxation.

This time, the condition is that $\dot{\varepsilon} = 0$. Then our differential equation reduces to

$$0 = \frac{\dot{\sigma}}{E} + \frac{\sigma}{\mu}.$$

Or, with some rearranging,

$$\dot{\sigma} + \left(\frac{E}{\mu}\right)\sigma = 0.$$

This equation has a solution in the form of

$$\sigma(t) = \sigma_0 e^{-Et/\mu},$$

which is why the stress decays exponentially. Also, from the initial stress (because of the spring) of $\sigma(0) = \sigma_0 = E\varepsilon$, we find the stress relaxation equation for the Maxwell model:

$$\sigma(t) = (E\varepsilon)e^{-Et/\mu}.$$

Kelvin (or Voigt) Model

The Kelvin (or Voigt) model is defined by a single spring element and a single dashpot element in parallel (see **figure 3.20**). It is most applicable for materials that have the properties both of elasticity and viscosity. The Kelvin model predicts creep more realistically than the Maxwell model, because in the infinite time limit the strain approaches a constant.

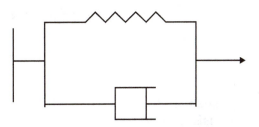

FIGURE 3.20   Schematic of a Kelvin (or Voigt) model consisting of a single spring element and a single dashpot element in *parallel*.

Although the Kelvin model is effective for predicting creep, it is not good at describing the relaxation behavior after the stress load is removed.

Again, the element equations can be used to derive the governing equation using what we know about composite models. Since the system is in parallel, the stresses are additive, such that

$$\sigma = \sigma_{spring} + \sigma_{dashpot}.$$

Now we can substitute the rules for our elements, $\sigma_{spring} = E\varepsilon_{spring}$ and $\sigma_{dashpot} = \mu\dot{\varepsilon}_{dashpot}$:

$$\sigma = E\varepsilon_{spring} + \mu\dot{\varepsilon}_{dashpot}.$$

Also, since the system is in parallel, the strains for the elements are equal, such that

$$\varepsilon = \varepsilon_{spring} = \varepsilon_{dashpot}.$$

Since we don't have to differentiate between the strains in the spring or dashpot, we can simplify our equation:

$$\sigma = E\varepsilon + \mu\dot{\varepsilon},$$

which is the governing differential equation for the Kelvin model.

1. For the Kelvin model, creep is characterized by the following (**figure 3.21**):

   b. An initial elastic strain of zero (because of the rigid dashpot at $t = 0$)
   c. Exponential increasing strain with time, approaching the elastic strain in the spring (because of the disappearing contribution of the dashpot)

We can consider our governing equation

$$\sigma_0 = E\varepsilon + \mu\dot{\varepsilon},$$

where "σ_0" is the initial applied (and constant) stress. The previous equation can be rewritten as

$$\dot{\varepsilon} + \left(\frac{E}{\mu}\right)\varepsilon = \frac{\sigma_0}{\mu}.$$

This formulation looks very similar to the creep equation for the Maxwell model, except

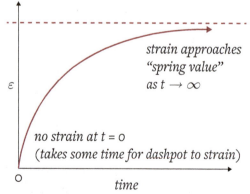

strain approaches "spring value" as $t \to \infty$

no strain at $t = 0$
(takes some time for dashpot to strain)

ε

time

O

FIGURE 3.21   Kelvin (or Voigt) model creep.

it is a differential equation dependent on strain, and it does not equal zero. The solution to this equation results in

$$\varepsilon(t) = \left(\frac{\sigma_0}{E}\right)(1 - e^{-Et/\mu}).$$

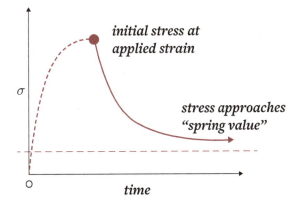

FIGURE 3.22   Kelvin (or Voigt) model relaxation.

For the Kelvin model, relaxation cannot be considered using an "instantaneously applied strain," as the dashpot would initially be rigid. Instead, we consider some previously achieved stress state, whereby we stop deforming a material and hold the strain constant. Then, relaxation is characterized by the following (**figure 3.22**):

a.   An initial stress owing to the current stresses in both the spring and dashpot
b.   An exponentially decreasing stress, as the dashpot relaxes, resulting in stress only because of the spring

The math for this situation is a little tricky, but the resulting equation for relaxation is

$$\sigma(t) = E\varepsilon_0 + (\sigma_0 - E\varepsilon_0)e^{-Et/\mu}.$$

Here, σ_0 represents the total initial stress in the spring and the dashpot at the time in which the strain is "held" such that "$\sigma_0 - E\varepsilon_0$" represents the stress in the dashpot alone.

III. Concluding Remarks

We began this chapter by considering simple models with springs of different stiffness arranged in parallel and in series, and we considered the force and displacement for each model. For springs in series, the force must be constant across springs, but the displacement is additive. For springs in parallel, the displacement must be constant across springs, and the force in each is additive for the total force.

Further, we observed that with composite materials (consisting of two or more distinct materials with unique properties), the different moduli can be combined similarly to combining spring stiffnesses by considering the materials of differing stiffness to be in series or parallel.

We have also learned that not all materials behave elastically (i.e., with an instantaneous response to load resulting in deformation). Some materials also exhibit viscous (i.e., time-dependent) behavior, where deformation occurs over time due to loading. Combined elastic and viscous behavior can be modeled by using simple model elements including springs (linear elastic) and dashpots (viscous). These model elements may be combined in series and in parallel to model material behavior. These types of models can be used to account for material creep (i.e., ongoing deformation under constant load) or relaxation (i.e., reduction in load over time due to material deformation). The two principal models we considered were (1) the Maxwell model—spring and dashpot in series; and (2) the Kelvin (or Voigt) model—spring and dashpot in parallel.

We next consider the chemical composition and structure of materials and how material chemistry at the atomic level influences physical and mechanical material properties.

IV. Problems

1. We talk about using springs in this chapter. What material property do the springs represent?

2. Name an example of a viscous solid.

3. Based on the behavior of springs and dashpots, explain the following:
    a. How the Kelvin model (spring and dashpot in parallel) behaves under constant applied stress over time.
    b. How the Maxwell model behaves under a constant applied strain over time.

4. Choose the model you think can best describe the behavior of glass and explain.

5. In a materials context, describe creep and name a material that may exhibit this phenomenon.

6. The figure shows a reinforced concrete column loaded axially by load, P. The cross-section shows that the overall diameter of the column is 20 inches, and there are 8 steel reinforcing bars, each 1 inch in diameter. The materials have the following properties:

Concrete:
Yield strength = 4,000 psi
$E = 1.2 \times 10^6$ psi
Steel:
Yield strength = 36,000 psi
$E = 30 \times 10^6$ psi

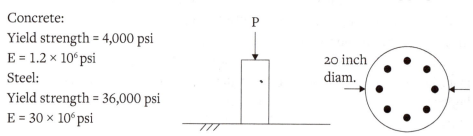

What is the maximum load, P, that can be applied to the column before it yields? (Hint: Are these materials in parallel, or in series?)

7. For the model shown, E_1 = 4000 kPa and E_2 = 20,000 kPa, answer the following:

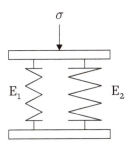

    a. If the model is compressed by a strain of 0.6 percent, what will be the stress in each spring due to this strain?

    b. If σ = 400 kPa is applied to the top plate, what strain will result?

8. For the Kelvin model shown, an initial strain, ε_0 = 0.017, is applied to the system using an initial stress, σ_0 = 10,420 psi, and the strain level is held constant.

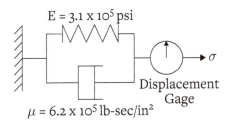

Using a spreadsheet, compute and plot the stress decay with time for this situation (your solution should show the equation you used for calculating the stress decay with time). Use time intervals of 0.1, 0.2, 0.3, 0.4, 0.5, 1, 2, 3 ... 15 seconds.

9. A 42-centimeter-long, 3-centimeter-diameter rod is composed of steel and aluminum, as shown in the figure. If the rod is loaded in tension by a 20 kN force, what is its total elongation, in millimeters? E_{steel} = 200 GPa and $E_{aluminum}$ = 69 GPa.

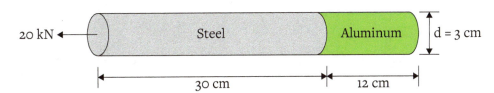

10. A 2.0-foot-long composite metal bar with a square cross-section is composed of a cylindrical copper core surrounded by steel. The core is 1.00 inch in diameter, and the square cross-section of the bar is 2.00 inches by 2.00 inches. Under a compressive load, the bar shortens by 0.01 inch. If E_{steel} = 29,000 ksi and E_{copper} = 17,000 ksi, what compressive force, in kips, is exerted on the bar to cause this deformation?

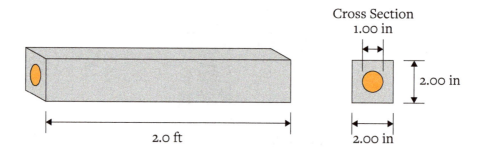

4

Material Chemistry

I. What Are Materials Made Of?

One of the classical mysteries of science was the question of what matter was ultimately composed of. If one considered all matter as made up of building blocks, what would happen if one were to divide those blocks into smaller and smaller pieces? Could such a process continue? Clearly, anyone who has taken an introductory chemistry class realizes the process can only reach as far as individual atoms.[1] In the Western tradition, Greek philosophers were the first to propose the existence of atoms; atoms were indestructible and immutable, and there are an infinite variety of shapes and sizes. They move through the void, bouncing off each other, sometimes becoming hooked with one or more others to form a cluster or molecule. Molecules of different shapes, arrangements, and positions give rise to the various macroscopic substances in the world. In chemistry, the history of molecular theory traces the origins of the concept or idea of the existence of strong chemical bonds between two or more atoms; the character and properties of these bonds give rise to material properties.

Consider the following scenario between an engineering student (who hasn't taken a course in materials) and a structural engineer looking at a recently constructed skyscraper:

The naïve student asks, "What is the building made of?"

And the engineer responds, "A perfectly reasonable question for a future engineer! Of course, the answer would be reinforced concrete."

The student, not yet familiar with reinforced concrete, replies, "But what is the reinforced concrete made of?"

---

1   Neglecting the high-energy physics involved with nuclear *fission,* of course.

Anticipating this question, the engineer explains, "You should use your head and think a little: reinforced concrete is simply composed of concrete reinforced with steel—so it is simply made of concrete and steel."

Being inquisitive, the student continues, "That makes sense, but what are concrete and steel composed of?"

The engineer, remembering his civil engineering materials class, easily answers, "In simplest terms, concrete is a mixture of cement, water, and aggregates,[2] while steel is an alloy of iron and carbon.[3]"

Truly intrigued now, the student continues, "I know from chemistry that iron and carbon are on the periodic table, but what are cement, water, and aggregates made of?"

The structural engineer, not being a chemist or materials scientist, thought for a short while and then recalled from his schooling, "Water is simply H_2O, or hydrogen and oxygen. Aggregates, such as crushed granite and sand, are mostly composed of silica, or SiO_2, or silicon and oxygen. Cement is a little more complicated, but I believe it is mostly composed of calcium silicates and aluminates, made up of calcium, oxygen, silicon, and aluminum."

Thinking a while, the student responds, "So the building is made of hydrogen, oxygen, silicon, calcium, aluminum, iron, and carbon? That doesn't seem like a lot of building materials."

The structural engineer replies, "The tricky part is knowing how to put the pieces together."

The same discussion can be focused on any structure or man-made system (such as a car or a cellular phone), and one will find out that almost everything is made up of a set of five to fifteen or so elements found on the periodic table. Even the most complex instruments may only have 20 elements or so. The key point is that there are a finite set of elements (currently 115) to construct everything—from skyscrapers to cellular phones, to spider webs, to bones.

That being said, there are literally thousands upon thousands of different materials because of the numerous different ways elements can interact and bond to form molecules, and how these molecules then interact to form larger-scale materials. Assembling a material is similar to construction of a building structure—instead of beams, there are molecules; instead of bolts, there are bonds. Like the engineer said, "The tricky part is knowing how to put the pieces together."

---

2   Covered in **chapter 6** "Aggregates and Cementitious Materials."

3   Covered in **chapter 5** "Metals and Steel."

Chemists and materials scientists discover the rules for material interactions and uncover new materials with extraordinary properties. Understanding the basic features at the atomistic level (i.e., the underlying chemical principles at the nanoscale) enables a better understanding of material properties and performance at the engineering scale.

II. Atomistic Interactions

If the basis of all materials is their constitutive atoms, then it follows that the behavior of materials is fundamentally dependent on the interactions between atoms. Just like the performance of a building can be linked to its structural components (e.g., beams and columns) and how they transfer loads, the performance of a material can be linked to its chemical structure and interactions. Theoretically, if we understand the atomistic interactions, we can predict the behavior of the material!

However, this task—understanding atomistic interactions—has been a great challenge for modern science. Theoretical physicists and chemists alike have toiled at the problem and have not yet fully solved it. Seemingly, with every advance, there is a new set of unsolved problems.

The basis of our current understanding of how atoms behave and interact is **quantum mechanics**. Quantum mechanics provides a mathematical description of the behavior and interactions of energy and matter, which have dual *particle-like* and *wavelike* properties. The mathematical formulations of quantum mechanics are abstract. In simple terms, a mathematical function known as the *wave function* provides information about the probability amplitude of position, momentum, and other physical properties of a particle. Perhaps the most popular expression is the Schrödinger equation:

$$i\hbar \frac{\partial}{\partial t} \psi(r,t) = \hat{H} \psi(r,t),$$

where i is the imaginary unit, the symbol "$\partial/\partial t$" indicates a partial derivative with respect to time t, $\hbar$ is Planck's constant divided by 2π, $\psi(r,t)$ is the wave function of the quantum system, and $\hat{H}$ is a Hamiltonian operator that characterizes the total energy of any given wave function and takes different forms depending on the situation. In effect, the Schrödinger equation is a partial differential equation that describes how the quantum state of some physical system changes with time (not unlike Newton's laws of motion). The Schrödinger equation can then be solved numerically for the electrons that orbit the nucleus of an atom and how they interact in the presence of other atoms.

The tricky part is that the wave function only gives the *probability* of the position and momentum of an electron; it does not give its precise information! Even worse, the more

FIGURE 4.1 Example atomic orbital or electron orbital. The orbitals describe a given space in which the probability of an electron may be found, defined by an electron's energy (n), angular momentum (ℓ), and vector (m). Specifically, here, $n = 4$, $\ell = 2$, and $m - 0$ (d orbital). The colors show the wave-function phase.

precisely the position of some particle is determined, the less precisely its momentum can be known, and vice versa. This is commonly known as **Heisenberg's uncertainty principle**.

As engineers, we don't like uncertainty. We prefer our equations to result in definitive numbers and solutions and our analytical approaches to have deterministic outcomes as much as reasonably possible.[4] For macroscopic engineering, we can rely on the steady shoulders of Newton rather than the uncertainties of Schrödinger, Heisenberg, and others. While the implications of quantum mechanics and the uncertainty principle are beyond the scope of this chapter and text in general, they are simply introduced to rationalize our chemical foundation of materials—specifically, the interactions between electrons.

Electron Configuration

The current atomistic model is one of a dense nucleus (begetting mass) surrounded by a cloud of electrons, which travel in atomic orbitals (**figure 4.1**):

The atomic orbital is a mathematical function that describes the wavelike behavior of either one electron or a pair of electrons in an atom. Each orbital in an atom is characterized by a unique set of values of the three quantum numbers n, ℓ, and m, which correspond to the electron's energy, angular momentum, and an angular momentum vector component, respectively. Any orbital can be occupied by a maximum of two electrons. The simple names s orbital, p orbital, d orbital, and f orbital refer to orbitals with angular momentum quantum number $\ell = 0$, 1, 2, and 3, respectively. These names, together with the value of n, are used to describe the electron configurations. Atomic orbitals are the basic building blocks of the atomic orbital model (alternatively known as the electron cloud or wave mechanics model). The repeating periodicity of elements within sections of the periodic table arise naturally from the total number of electrons that occupy a complete set of s, p, d, and f atomic orbitals, respectively.

---

4   Uncertainty is a fact of life that impacts many aspects of civil engineering; for example, there is uncertainty associated with the occurrence of earthquakes and extreme weather events, as well as the quality of materials and construction. It is important that engineers can (and do) account for uncertainty on a regular basis.

Ignoring the quantum mechanical interpretations in favor of more deterministic approaches, we can assume the following:

1. Protons/neutrons form the nucleus at the center of an atom, and electrons travel around the nucleus in a shell or path (atomic orbital).
2. Each shell or path holds, or can hold, only a fixed number of electrons.
3. Atoms desire a minimum energy level, accomplished by filling shells with electrons or simulating this with shared charges; electrons with too much energy "jump" to adjacent shells.
4. Electrons in the outermost shell are known as valence electrons; these electrons are responsible for bonding, as they are most likely to be imbalanced and/or share the orbitals with adjacent atoms.

When electron orbitals interact or combine (e.g., an electron effectively orbits two nuclei), and this interaction/combination results in a decrease in energy, then the atoms are said to **bond**.

Bonding Types

A chemical bond is an attraction between atoms that allows the formation of chemical substances that contain two or more atoms. Ultimately, the bond is caused by the electrostatic force of attraction between opposite charges—either directly through the negative charge of the electrons and the positive charge of the nuclei or indirectly as the result of a dipole attraction.

Simply put, opposite charges attract; via a simple electromagnetic force, the negatively charged electrons that are orbiting the nucleus of an atom, and the positively charged protons in the nucleus of an atom, attract each other. An electron positioned between two nuclei will be attracted to both of them, and the nuclei will be attracted toward electrons in this position. This attraction constitutes the chemical bond.

In general, strong chemical bonding is associated with the sharing or transfer of electrons between the participating atoms. The atoms in molecules, crystals, metals, and diatomic gases—indeed the entirety of the physical environment around us—are held together by chemical bonds, which dictate the structure and the bulk properties of matter and materials.

The primary reason for chemical bonding is atoms trying to achieve lower energy states. You can think of it as attempting to "fill" electron orbitals, which have the lowest energy only when they are full (this is why, in general, atoms interact proportionally to the change in number of electrons in outer shells). Consider the example of hydrogen, H, which only has a single electron in the smallest (s) orbital. The ground state energy of a hydrogen atom is approximately –13.6 eV. Coming across another H atom with a single electron, the electrons are attracted to the orbital/nucleus of each other (there is mutual room for an extra electron), and thus a bond is formed:

$$H + H \rightarrow H_2,$$

resulting in a two-atom hydrogen molecule. One might suspect that the ground state energy of the hydrogen molecule is –27.2 eV (i.e., twice the value of a single H atom), but this is not the case. The ground state energy of a hydrogen molecule is approximately –32.0 eV! The formation of the bond (and filling the orbitals of two atoms) has reduced the total energy by 4.8 eV. That is why the bond forms: because the energy decreases!

The bond in the hydrogen molecule is defined as a covalent bond. We will now proceed to look at the most common types of chemical bonds.

Covalent Bonds

A covalent bond involves the sharing of electron pairs between atoms (e.g., when two or more electrons are shared by elements to fill orbitals). From an energy perspective, atoms "like" to have the electron structure of a noble gas. The result is a stable balance of attractive and repulsive forces between atoms. For many molecules, the sharing of electrons allows each atom to attain the equivalent of a full outer shell, corresponding to a stable electronic configuration (similar to the closest noble gas on the periodic table), and thus covalent bonding commonly occurs among electronegative elements, such as carbon and oxygen.

Examples:

- **Hydrogen:** Each H atom only has a single electron. Thus, it likes to share with other H atoms such that the electrons are paired, and a single covalent bond is formed (**figure 4.2**).

- **Methane:** Each carbon atom has four electrons in its outermost shell. Thus, it requires four more electrons to acquire a stable noble gas configuration (Ne configuration). Each of the hydrogen atoms has only one electron in its outermost shell and requires one more electron to complete its outermost shell (to acquire He configuration). To achieve this, one carbon atom forms four single covalent bonds with four hydrogen atoms (**figure 4.3**).

- **Carbon dioxide:** Each carbon atom has four electrons in its outermost shell, and each oxygen atom has six electrons in its outermost shell. Thus, each carbon atom requires four, and each oxygen atom requires two more electrons to acquire noble gas configurations (Ne).

FIGURE 4.2 Graphical representation of two hydrogen atoms coming together and sharing their single electrons to create a full outer-shell pairing.

To achieve this, two oxygen atoms form a double covalent bond with carbon; each oxygen shares two electrons with the carbon atom (**figure** 4.4).

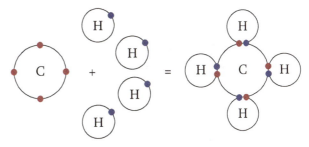

FIGURE 4.3 Graphical representation of methane. Four hydrogen atoms and one carbon atom come together and share electrons to create a full outer shell.

Ionic Bonds

Ionic bonding is a type of chemical bonding that involves the electrostatic attraction between oppositely charged ions. These ions are atoms that have lost one or more electrons (and thus have a net positive charge and are known as **cations**) or atoms that have gained one or more electrons (and thus have a net negative charge and are known as **anions**). In the simplest case, the cation is a metal atom (elements in groups I and II of the periodic table; e.g., K^+), and the anion is a nonmetal atom (elements in groups VI and VII; e.g., Cl^-), but ions can also be of a more complex nature (e.g., molecular ions like NH_4^+ or SO_4^{2-}).

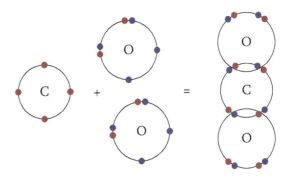

FIGURE 4.4 Graphical representation of carbon dioxide. Two oxygen atoms and one carbon atom come together and share electrons to create a full outer shell.

It is important to note that both covalent and ionic bonding involve unbalanced electron orbitals. Moreover, pure ionic bonding—in which one atom completely "steals" an electron from another—cannot exist: all ionic bonds and compounds have some degree of covalent bonding or electron sharing. The term "ionic bonding" is given when the ionic character is greater than the covalent character (when the bonding is to be more polar (ionic) than in covalent bonding, where electrons are shared more equally).

Examples:

- **Table salt (sodium chloride)**: Sodium chloride, also known as common table salt, is an ionic compound with the formula NaCl, representing equal proportions of sodium and chlorine. Sodium has a single electron in the outermost orbital and thus is a highly reactive alkali metal. Losing an electron, Na becomes a positively charged cation (Na^+). Chlorine requires one electron in the outermost orbital and has a high electron affinity. The "extra" electron from Na can be predominately taken by Cl. With the extra electron, Cl becomes a negatively charged anion (Cl^-). The two atoms (Na and

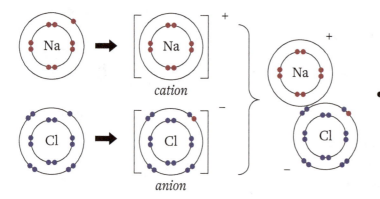

FIGURE 4.5   Sodium chloride. Sodium loses an electron from its outermost orbital to become a cation, while chlorine acquires an electron in its outermost orbital, becoming an anion. Being oppositely charged, sodium and chlorine attract each other to form an ionic bond.

Cl) are now oppositely charged, so they attract each other and form an ionic bond (**figure 4.5**).

- **Magnesium oxide:** An ionic compound consisting of a lattice of Mg^{2+} ions and O^{2-} ions held together by ionic bonding. It is similar to NaCl, but with two electrons being exchanged (note Mg is next to Na on the periodic table, and O is next to Cl). Losing two electrons, Mg becomes a positively charged cation (Mg^{2+}). Likewise, with two extra electrons, O becomes a negatively charged anion (O^{2-}). The two atoms (Mg and O) are now oppositely charged, so they attract each other and form an ionic bond (**figure 4.6**).

- **Ionic solids:** Formed because of the attraction between positive and negative ions, typically in a fixed ion arrangement. Many salt molecules (NaCl), for example, combine to form a salt crystal, as shown in **figure 4.7**.

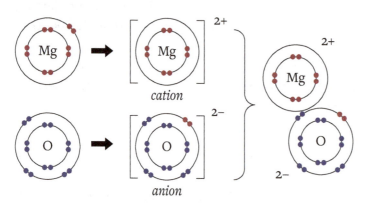

FIGURE 4.6   Magnesium oxide. Magnesium loses two electrons from its outermost orbital to become a cation of (+2) charge, while oxygen acquires two electrons in its outermost orbital, becoming an anion of (–2) charge. Being oppositely charged, magnesium and oxygen attract each other to form an ionic bond.

Here lies the main difference between ionic bonding and covalent bonding. For covalent bonding, the atoms share electrons with each other, and thus the bond is one-to-one and directional. For example, in methane, each hydrogen atom requires one more electron, which is provided by a carbon. In effect, the hydrogen has "married" that carbon—they are stuck together for life![5] Moreover, the four hydrogens involved in methane

---

5   A chemical reaction can be considered an atomistic "divorce."

will always maintain the same relative directions to each other (forming a tetrahedral shape). Now, compare that to the structure of salt noted earlier. Consider one of the sodium atoms: How many chlorines is it "bonded" to? In the 2-D structure, an interior Na has four Cl neighbors, even though it is only charged with one lost electron. An ionic bond is neither one-to-one nor directional—it will attract as many opposite charges as it can until the space around it is filled. If a covalent bond is considered a marriage between two atoms, ionic bonds can be considered dating singles: they couple when it is convenient, but if someone more attractive comes along, then goodbye!

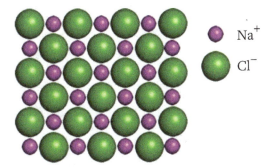

FIGURE 4.7   Ionic solids, formed from attraction between positive and negative ions. Typically, the ions assemble in a fixed arrangement. For salts (like sodium chloride) the ion arrangements form salt crystals.

Metallic Bonds

Metallic bonds typically consist of electropositive elements that want to lose electrons but have "too many" to give such that there are no convenient pairings for a covalent bond. For example, aluminum has three electrons in its outer shell. It would need five extra electrons to fill the shell or lose three to satisfy its orbitals. This is energetically unfavorable.[6] Aluminum (and other metals) effectively share their excess electrons such that all atoms can "fill their shells" with the common pool.

Metallic bonds can be considered socialist bonds—they share their belongings for the common good!

As depicted earlier, in metallic bonds, atoms give up electrons to a common "pool" that are not associated with any particular atom. Electrons are "free agents," moving around an assembly of atoms, and all the atoms "satisfy" their orbital requirements, producing an energetically and electrically stable configuration. As a result of these "free electrons," metals typically have high electrical and thermal conductivity (**figure 4.8**).

Secondary Bonds

Secondary bonds are weaker molecular interactions, but are still important for material properties. They include the following:

- **Polar interactions**: Internal "imbalance" in atomic charge (more negative on one end than the other). Atoms attracted by electrostatic force (dipole attraction). A dipole occurs when electrons stay closer to one element in a molecule/bond (e.g., water).

---

6   But not impossible. Aluminum can exist as an Al^{3+} ion.

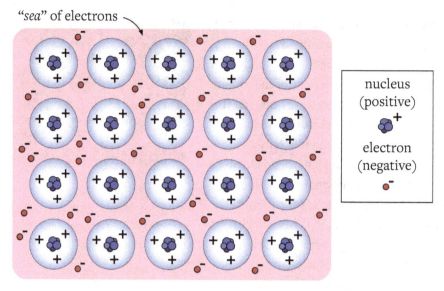

"*sea*" of electrons

nucleus
(positive)

electron
(negative)

FIGURE 4.8 In metallic bonds, metal atoms give up electrons to a common "pool" around the metal atoms. These free electrons result in metals typically having high electrical and thermal conductivity.

- **Van der Waals bonding**: Weak attractive forces between all atoms/molecules, also related to dipole interactions (e.g., graphite).
- **Hydrogen bonds** (**H-bonds**): A hydrogen bond is the electrostatic attraction between polar molecules that occurs when a hydrogen (H) atom is attracted to a highly electronegative atom, such as nitrogen (N), oxygen (O), or fluorine (F). The hydrogen bond is stronger than a van der Waals interaction, but weaker than covalent or ionic bonds. This type of bond can occur in inorganic molecules, such as water, and in organic molecules, such as DNA and proteins. The hydrogen bond is often described as an electrostatic dipole-dipole interaction. However, it also has some features of covalent bonding: it is directional and strong, produces interatomic distances shorter than the sum of the van der Waals radii, and usually involves a limited number of interaction partners (e.g., DNA).
- **Electrostatic interactions**: Attractions between charged ions that cause attraction, but prevent them from causing ionic bonds (e.g., polyelectrolytes).

III. Quantifying/Modeling Bond Energies

The forces holding a solid together can be considered the simple sum of attractive forces plus repulsive forces. Measurements and quantum mechanical calculations have shown that atoms attract each other via a decaying function with respect to distance. This attraction can be approximated by

$$U_{attraction} = -N\frac{A(z_1 \cdot z_2)e^2}{r},$$

where N is the number of atoms, A is a constant (a function of the atomic structure; e.g., free electrons and electron configuration), z_1 and z_2 the charges of adjacent atoms/molecules, e the fundamental electric charge, and r the interparticle or interatomic distance.

Atoms repel each other at short ranges because of Pauli repulsion (overlapping electron orbitals). This repulsion is difficult to theoretically deduce, but is typically represented by a general expression:

$$U_{repulsion} = N\frac{B}{r^a},$$

where, again, r is the distance between atoms, N the number of atoms, and B and a are fitted constants.

Adding the functions, the total atomistic energy is

$$U = U_{total} = U_{attraction} + U_{repulsion} = -N\frac{A(z_1 \cdot z_2)e^2}{r} + N\frac{B}{r^a}.$$

This expression is most stable when $r = r_0$, $U = U_{min}$. This is a relatively simple minimization problem that can be solved for

$$\frac{\partial U}{\partial r} = 0.$$

Of note is that, from a molecular perspective, compressing atomistic bonds always results in an energy increase; there is no "bond breaking" in compression. As a result, no material technically "fails" in compression. Stress can be increased in compression until the bond energy is so high that the atoms "slip" in a different direction. This results in a shear failure (which we will discuss later).

Atomistic Potentials

The earlier formulation is just one approximation. Any function that reflects the attractive and repulsive forces can be used, and many types of equations are fitted to quantum mechanical data. We will look at two common functions, also called atomistic potentials:

1. The Lennard-Jones potential
2. The Morse potential

1. Lennard-Jones Potential

The **Lennard-Jones potential** (also referred to as the **L-J potential**, **six to twelve potential**, or **twelve-to-six potential**) is a mathematically simple model that approximates the interaction between a pair of neutral atoms or molecules. A form of the potential was first proposed in 1924 by John Lennard-Jones. The most common expressions of the L-J potential are

$$U_{LJ} = 4\varepsilon \left[\left(\frac{\sigma}{r} \right)^{12} - \left(\frac{\sigma}{r} \right)^{6} \right],$$

where ε is the depth of the potential well, σ is the finite distance at which the interparticle potential is zero, r is the distance between the particles, and r_m is the distance at which the potential reaches its minimum. At r_m, the potential function has the value $-\varepsilon$. The distances are related as $r_m = 2^{1/6}\sigma$. These parameters can be fitted to reproduce experimental data or accurate quantum chemistry calculations. The L-J potential is a relatively good approximation, and because of its computational simplicity, it is used extensively in computer simulations, even though more accurate potentials exist.

The L-J potential is often used to describe the properties of gases, and to model dispersion and overlap interactions in molecular models. It is particularly accurate for **noble gas** atoms and is a good approximation at long and short distances for neutral atoms and molecules. The sixth-power term arises as a result of dipole-dipole interactions because of electron dispersion in noble gases, but it does not represent other kinds of bonding well. The twelfth-power term appearing in the potential is chosen for its ease of calculation for simulations (by squaring the sixth-power term) and is not physically based.

2. Morse Potential

The **Morse potential**, named after physicist Philip M. Morse, is another convenient model for the potential energy of a diatomic molecule. It also accounts for the anharmonicity of real bonds. The Morse potential can also be used to model other interactions, such as the interaction between an atom and a surface. Because of its simplicity (only three fitting parameters), it is commonly used in simulations. The Morse potential energy function is of the form

$$U(r) = \varepsilon \left(1 - e^{-a(r-r_0)} \right)^{2},$$

where r is the distance between the atoms, r_0 is the equilibrium bond distance, ε is the well depth (defined relative to the dissociated atoms), and a controls the "width" of the potential (the smaller a is, the larger the well).

Of interest, the stiffness of the bond, k, can be found by Taylor expansion of $U(r)$ around r_0 to the second derivative of the potential energy function, from which it can be shown that the parameter, a, is related, where

$$a = \sqrt{k/2\varepsilon},$$

where k is the equivalent spring stiffness constant at the minimum of the well.

Finally, since the zero of potential energy is arbitrary, the equation for the Morse potential can be rewritten any number of ways by adding or subtracting a constant value. When it is used to model the atom-surface interaction, the energy zero can be redefined so that the Morse potential becomes

$$U(r) = \varepsilon \left(e^{-2a(r-r_0)} - 2e^{-a(r-r_0))} \right)^2,$$

where r is now the coordinate perpendicular to the surface. This form approaches zero at infinite r and equals $-\varepsilon$ at its minimum (i.e., $r = r_0$). It clearly shows that the Morse potential is the combination of a short-range repulsion term and a long-range attractive term, analogous to the Lennard-Jones potential.

Thermal Expansion

We learned in **chapter 2** that materials expand and contract when subject to changes in temperature. This phenomenon occurs because when atoms heat up, the increase in *kinetic energy* allows them to "expand" as they vibrate about their bonds.

IV. Metallic Materials

Due to metallic bonding, most metals undertake a particular crystalline (ordered) arrangement, which is known as the **lattice structure**. This structure defines properties and behaviors unique to metals, such as defects, **dislocations**, and **grains**.

Lattice Structure(s)

A lattice structure is simply a patterned structure that is spatially repeated (typically in three dimensions for metals). The structure itself can then be represented by a "unit cell" that represents the features of the structure—the solid can then be thought of as a combination of these repeated unit cells.

The arrangement of atoms is represented by so-called **space lattices** (**figure 4.9**), of which the most common types for metals are as follows:

1. FCC: Face-centered cubic
2. BCC: Body-centered cubic
3. HCP: Hexagonal close-packed

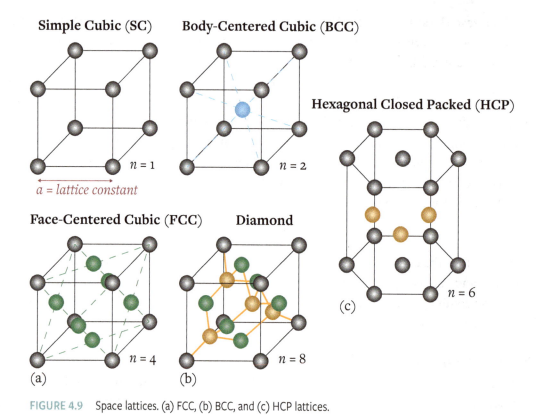

Simple Cubic (SC)

$n = 1$

$a = lattice\ constant$

Body-Centered Cubic (BCC)

$n = 2$

Hexagonal Closed Packed (HCP)

$n = 6$

(c)

Face-Centered Cubic (FCC)

$n = 4$

(a)

Diamond

$n = 8$

(b)

FIGURE 4.9 Space lattices. (a) FCC, (b) BCC, and (c) HCP lattices.

How atoms are packed together varies how many bonds (neighbors) an individual atom has, how dense the material is, and the potential for alloying (adding additional elements).

Atomic Packing Factors

The atomic packing factor (APF), or **packing fraction**, is the fraction of volume in a crystal structure that is occupied by atoms. It is dimensionless and always less than unity. For practical purposes, the APF of a crystal structure is determined by assuming that atoms are rigid spheres, and, for one-component crystals (those that contain only one type of atom), the APF can be represented mathematically by

$$APF = \frac{n \times V_{atom}}{V_{unit}},$$

where n is the number of atoms in the unit cell, V_{atom}, the spherical volume of a single atom, and V_{unit}, the volume of the unit cell itself.

For cubic crystal systems, only the atomic diameter is required to calculate the APF. Consider BCC crystals. From the geometry of the atoms, we can relate the diameter (D) of the atom to the lattice constant, a, where

$$(2D)^2 = a^2 + 2a^2$$

or

$$D = \frac{\sqrt{3}}{2}a.$$

Rearranging slightly,

$$a = \frac{2D}{\sqrt{3}},$$

from which we can calculate the volume of the unit cell:

$$V_{unit} = a^3 = \left(\frac{2D}{\sqrt{3}}\right)^3 = \frac{8D^3}{3\sqrt{3}}.$$

We also know the volume of a single atom:

$$V_{atom} = \frac{4}{3}\pi r^3 = \frac{4}{24}\pi D^3 = \frac{\pi D^3}{6}.$$

Finally, for BCC, we know that there are two atoms per unit cell (one center atom, four corner atoms), or n = 2. Thus, the APF is

$$APF = \frac{n \times V_{atom}}{V_{unit}} = 2 \times \frac{\pi D^3}{6} \times \frac{3\sqrt{3}}{8D^3} = \frac{\pi\sqrt{3}}{8} \cong 0.68 .$$

By similar procedures, the ideal APFs of common crystal structures can be found (see **table 4.1**).

TABLE 4.1 ATOMIC PACKING FACTORS OF COMMON CRYSTAL STRUCTURES

Crystal Structure	APF
HCP	0.74
Simple cubic	0.52
FCC	0.74
BCC	0.68
Diamond cubic	0.34

Of interest is the concept of **random close-packing**—the maximum volume fraction of solid objects obtained when they are packed randomly. For example, when a solid container is filled with aggregate (e.g., crushed stone), shaking the container will reduce the volume taken up by the objects, thus allowing more aggregate to be added to the container. In other words, shaking increases the density of packed objects. Experiments have shown that the most compact way to pack *spheres* randomly gives a maximum density of about 64 percent. This is significantly smaller than the maximum theoretical filling fraction of 0.74 that results from HCP (also known as close-packing).

Lattice Defects

Ideal metal has a perfect lattice structure, but defects exist, including the following:

- Missing atoms
- Line defects or rows of displaced/missing atoms—dislocations
- Grain boundaries
- Cavities in the material

Of key importance is the presence of dislocations. Dislocations occur when shear stress is applied and line of atoms displaced progressively. When stress is released, the defect gets "stuck" in the material (**figure 4.10**). Dislocations are the cause of plasticity in metals.

FIGURE 4.10   Example of dislocation motion.

Grain Structure

Viewed microscopically, metals exhibit grain-like structures on their surface. These "grains" are areas where the crystal lattice structures are oriented in different directions. In effect,

- metals are produced by heating ingredients and then cooling them to form crystals;
- multiple crystal "zones" or nuclei form with different alignments;
- where those alignments meet are called grain boundaries;
- grain boundaries help the strength of material by blocking slip or weakness planes (i.e., dislocations cannot move past grain boundaries);
- therefore, it takes more energy to fail the material, making it stronger;
- however, since dislocations are prevented from moving, plasticity is lost, and the material becomes more brittle.

The smaller the grain size in the metal, the more grain boundaries that exist and the stronger the metal will be (to a point; once the grains get small enough, the mechanism of failure transitions to grain boundary sliding instead of dislocation movement, and the metal starts getting weaker). The association between strength and grain size is known as the Hall-Petch relationship, which is an inverse relationship.

Alloys

An **alloy** is a mixture of two or more elements in which the main component is a metal. Most pure metals are either too soft, brittle, or chemically reactive for practical use. Combining different ratios of metals as alloys modifies the properties of pure metals to produce desirable characteristics. The aim of making alloys is generally to make materials that are less brittle, harder, more resistant to corrosion, or have a more desirable color and luster. Of all the metallic alloys in use today, the alloys of iron (steel, stainless steel, cast iron, tool steel, alloy steel) make up the largest proportion both by quantity and commercial value. Iron alloyed with various proportions of carbon gives low-, mid-, and high-carbon steels, with increasing carbon levels reducing ductility and toughness. The addition of silicon will produce cast irons, while the addition of chromium, nickel, and molybdenum to carbon steels (more than 10 percent) results in stainless steels.

Other significant metallic alloys are those of aluminum, titanium, copper, and magnesium. Copper alloys have been known since prehistory—bronze gave the Bronze Age its name—and have many applications today, most importantly in electrical wiring. The alloys of aluminum, titanium, and magnesium are valued for their high strength-to-weight ratios; magnesium can also provide electromagnetic shielding. These materials are ideal for situations where high strength-to-weight ratio is more important than material cost, such as in aerospace and some automotive applications.

> **Example: Aluminum alloy = 5.6 percent zinc, 2.5 percent magnesium, 1.6 percent copper, 0.3 percent chromium**

When alloying, the "extra" element is placed either in voids of the lattice (called interstitial atoms) or replaces atoms in the lattice (called substitutional atoms). Alloying elements can also help block dislocation movement and thus strengthen the material.

Alloy Phase Diagrams

To create an essential building component, such as steel, we must alloy iron with various other elements, such as carbon, manganese, nickel, chromium, and boron. The process of mixing these ingredients is not as simple as combining eggs, sugar, and flour with heat to create a cake. To create a metallic alloy, all ingredients must first be heated to a molten state (liquid-like). It is then that they can be mixed to form the appropriate alloy and cooled to achieve the desired performance properties. What is most important in this whole process is controlling temperature. The temperature at which materials transition between liquid and solid changes as we mix the different materials together; however, this behavior is predictable. Material scientists and metallurgists have created plots called **alloy phase diagrams** that graphically help in calculating the amount of solid and liquid in an alloy at specific temperatures based on the percentages of the individual components. A simplified two-element alloy phase diagram is provided in **figure 4.11**.

Most phase diagrams will have temperature on the y-axis and percentage of the materials on the x-axis. Note that the top and bottom axes of this graph are for the percentage of each of the two materials in the alloy. Where one starts at zero, the other begins at one hundred, which simply means that if you start with 100 percent of one element, then you have 0 percent of the other.

When you have 0 percent of the second element, then the melting point of that material is definite because only one material is present. At temperatures below the melting point, the material is a solid; at temperatures above that point, the material is a liquid. As you introduce another element (moving to the right on the plot in **figure 4.11** as increasing amounts of material B compose the alloy), the amount of solid and liquid forms of each material can vary at a given temperature.

For a given alloy blend, the temperature defines a **state point**, which represents the specific composition at a specific temperature. The upper line in the plot in **figure 4.11** is known as the **liquidus**, and the lower line is the **solidus**. If the temperature is at or above the intersection of an alloy blend with the liquidus line (i.e., the state point plots

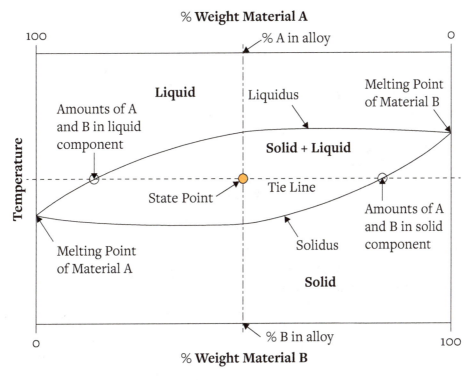

FIGURE 4.11   Two-element alloy phase diagram.

above the liquidus line), then the material will be all liquid. Likewise, if the temperature is at or below the intersection of an alloy blend with the solidus line (i.e., the state point plots below the solidus line), then the material will be all solid. If the temperature of an alloy blend is such that the state point falls between the liquidus and solidus lines, then both solid and liquid are present in the blend at the same time. The amount of liquid and solid for each element can be found by drawing a horizontal **tie line** at the temperature of interest (i.e., through the state point) that crosses both the liquidus and solidus lines. The intersection of the tie line with the liquidus line indicates the proportion of each material in the liquid state at the given temperature, and the intersection of the tie line with the solidus line indicates the proportion of each material in the solid state at the same given temperature.

V. Organic Solids

Presently, there are four types of organic solids used in civil engineering applications, with some more common than others. We will introduce these materials here in the context of

EXAMPLE 4.1: USING A PHASE DIAGRAM TO DETERMINE ELEMENT PHASE COMPONENTS

Antimony (Sb) and bismuth (Bi) are metals that when alloyed can be very useful for a variety of applications, from soldering to radiation shielding. Given the phase diagram for Bi-Sb (**figure 4.12**), determine the masses that are in liquid and solid phases at 450°C if the total mass of the alloy is 100 grams and you know that 40 percent of the alloy is antimony.

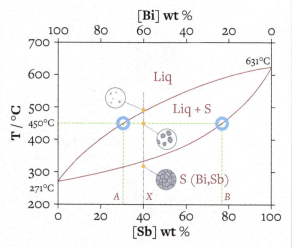

Figure 4.12   *Using a phase diagram to determine element phase components.*

SOLUTION

Since we know that 40 percent of the alloy is antimony, and the total mass is 100 grams, the distribution is thus

Sb = 40 g and Bi = 60 g.

From the diagram, we see that at 450°C we have both liquid and solid, and the following values can be found from the liquidus and solidus (indicated by the blue circles):

	Sb	Bi
Percent liquid	30	70
Percent solid	77	23

Next, we can determine the *mass* of the alloy in the respective states by writing two mass balance equations and solving simultaneously.

Sb: $[40 \text{ g} = 0.3(M_L) + 0.77(M_S)]$ $\times 7$ $\rightarrow$ $280 \text{ g} = 2.1(M_L) + 5.39(M_S)$

Bi: $[60 \text{ g} = 0.7(M_L) + 0.23(M_S)]$ $\times 3$ $\rightarrow$ $180 \text{ g} = 2.1(M_L) + 0.69(M_S)$

Subtracting Bi from Sb: $100 \text{ g} = 4.70(M_S)$

Solving for M_S: $M_S = 21.27 \text{ g}$

Solving for M_L: $M_L = 78.73 \text{ g}$

Therefore,

	Total	Liquid	Solid
Sb	40 g	$0.3(M_L) = 23.62 \text{ g}$	$0.77(M_S) = 16.38 \text{ g}$
Bi	60 g	$0.7(M_L) = 55.11 \text{ g}$	$0.23(M_S) = 4.89 \text{ g}$
Total	100 g	78.73 g	21.27 g

their chemical makeup and their associated behaviors. It is important to point out that the term "organic" means that the materials are carbon based. The carbon-based materials are formed by polymers, which are large molecules in repeating subunits. These long, irregular molecular chains can show big differences in chemical and mechanical behavior based on their intermolecular-bonding type.

Secondary Bonding and Cross-Linking

Secondary bonds are bonds between polymer chains. While primary bonds are classified as ionic, covalent, and metallic, secondary bonds are weaker bonds classified broadly as van der Waal's forces and hydrogen bonds. **Van der Waal's forces** can be found naturally where there is a meeting of two dipoles (molecules with oppositely charged ends, or polar molecules). It is said that geckos, spiders, and other insects use van der Waal's forces between their leg fibers and walls to help them climb up seemingly impossible surfaces with ease—a discovery that has resulted in much study as to how we can mimic this to create better adhesives. **Hydrogen bonds** are the stronger of the secondary bonds. They occur when hydrogen is covalently bound to oxygen or nitrogen. This bond creates a partially positive charge, which then attracts the hydrogen to the oxygen or nitrogen (which are electronegative) of another molecule to achieve balance. One common type of hydrogen bond is the attraction of water molecules to one another. You can actually see this effort at work when you place two water drops close to one another on a table: they will want to merge!

Cross-linking is when two or more molecules are joined with a covalent bond. Cross-links provide stability for the molecules, resulting in increased material strength, decreased strains, and an increased melting point. The addition of cross-links can even turn liquids into gels or solids, depending on the added concentration. The chemical process of vulcanization is a type of cross-linking achieved in tires by adding a curing agent (usually sulfur) to natural rubber and heating it under high pressure to encourage the cross-link bonding. The cross-links enable tires to withstand higher and lower temperatures without becoming stickier or brittle, respectively, as they would in just their natural form.

Common Organic Solids

Thermoplastics are a category of plastic materials most noted for their ability to become viscous when heated, making them highly recyclable. In fact, they can be reheated or softened repeatedly without chemical breakdown, which is important for distinguishing them from other plastics. The ability of this material to form and reform is due broadly to the intermolecular bonds, which employ van der Waals forces, to form linear polymer structures. These linear structures can be in one of two categories: amorphous (chains form bundles that look like a plate of spaghetti) and crystal (organized alternating layers of polymers), with the crystal structures dictating the mechanical properties (including stress, strain, and temperature resistance) of the material. Some drawbacks to their chemical makeup, with respect to engineering purposes, include deformation when heated and swelling or solubility in the presence of certain solvents. Some well-known thermoplastics include polyethylene (geomembranes), polypropylene (textiles), PVC (piping), asphalt (roads), and Teflon (catheter coating that prevents infectious agents from adhesion!)

Thermosets are plastics made with resins or hardeners known for their durability, resistance to thermal fatigue, high resistance to creep, and nonreactivity to solvents. These plastic "superpowers" are a result of their unique chemical structure, which employs linear structures similar to thermoplastics with added cross-links between the chains. These cross-links give the plastic added rigidity and allow it to have increased mechanical performance, with the caveat of brittle failure and the inability to be melted and reformed (recycled). Thermosets, instead, will decompose when they are subjected to extreme heat because of their cross-link structure. Common thermosets include epoxy adhesives, polyester (fabrics and fiberglass), and melamine (laminate flooring and work surfaces).

Elastomers are highly elastic materials that are characterized as having long polymer chains with the occasional cross-link (but not as many as a true thermoset). These long chains are normally coiled up but when stressed will uncoil and allow the elastomers to be stretched to lengths up to seven times their original size. Once released, elastomers can return to their original state. Well-known elastomers include natural and synthetic rubbers (gaskets), neoprene (wire insulator), and silicone (coating for salt and rain resistance).

Natural materials are a broad organic solid category, ranging from wood to protein-based materials such as spider webs. As sustainability becomes more important to how we design, engineers are looking more toward natural materials for inspiration and materials, a field called biomimicry. While wood does have its own chapter in this book, in the context of chemistry, it is pertinent to note that wood is approximately 40–50 percent cellulose. Cellulose is a highly linear chain of glucose held together with hydrogen bonds, making it resistant to tensile forces and thus a useful natural building material. Another natural material that is currently being studied is spider silk, which is consists of protein chains that are held together through hydrogen bonds, which give the silk its high tensile strength. In fact, some species of spider can produce silk with tensile strength that is comparable to steel! However, where this material really shines is its strength-to-density ratio, which far exceeds steel, making it a noteworthy lightweight material with sustainability implications.

VI. Concluding Remarks

What we have tried to make clear is that understanding the chemical makeup, the character and properties of bonds, helps us to predict a material's physical and mechanical properties. Why does stainless steel behave differently than cast iron? We now know that interstitial carbon plays a large role in resisting movement at the atomic level. Changes smaller than the eye can see produce results that can significantly improve how materials perform! Even the types of bonds between the atoms making up our materials can mean the difference between a flexible or brittle material and our ability to recycle a plastic or not. This understanding will be critical to the budding engineer entering into a career where the ability to weigh sustainability and performance is playing an increasing role.

VII. Problems

1. Describe the difference between a covalent bond and metallic bond.

2. Describe the difference between primary and secondary bonds.

3. Order the three primary bonding types discussed in this chapter by increasing conductivity. What contributes to this in each bond type?

4. Iron at 20°C is a body-centered cubic with atoms of radius 0.124 nanometers. Calculate the following:
   a. The side dimension of the lattice cube, "a". Hint: Draw a picture of this structure, assuming that the corner atoms are in contact with the body-centered atom.
   b. The maximum diameter of the atoms that can be placed interstitially between two corner atoms in the lattice structure.

5. What are the tradeoffs for cross-linking in plastics?

6. Why do engineers use alloys? Why not use pure metals?

7. What do Van der Waal's forces do to bonding energy?

8. Based on the plot of total atomistic energy, U_{net} versus the distance between atoms, r, shown in the figure, determine the most stable interatomic radius.

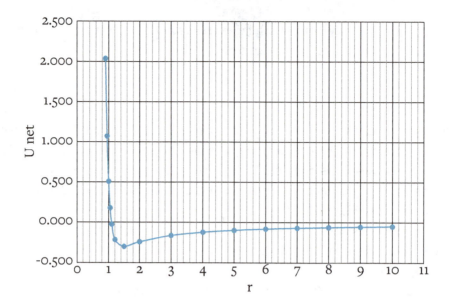

9. The figure that follows shows a phase diagram for two metals, A and B.

   a. What is the melting temperature of material A?

   b. For a temperature of 1,000°C, what proportion of the liquid is composed of material A and material B? What proportion of the solid is composed of material A and material B?

   c. An alloy of materials A and B is 75 percent material A and 25 percent material B. If you have 1,000 grams of the alloy at 1,000°C, how much of the material (by mass) is solid and how much is liquid?

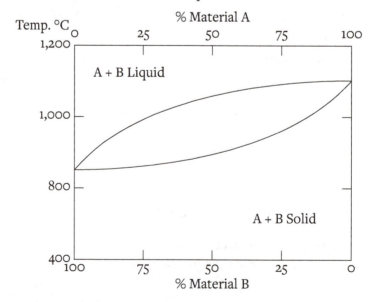

10. A new element has been shown to follow the following relation for interatomic energy (U as a function of r):

$$U = 4\varepsilon \left[\left(\frac{B}{r} \right)^2 - \frac{B}{r} \right].$$

Using quantum mechanics calculations, the atoms have a cohesive energy of $\varepsilon = 10$ eV at an equilibirium bond distance of $r_0 = 0.6$ nanometers.

Solve for the constant "B" such that the atomistic energy is minimized at r_0.

Credits

5

Metals and Steel

I. Introduction—Metals Driving Civilization and Technology

While concrete and wood have been implemented as construction materials for arguably as long, no other material type has motivated innovation and driving the advancement of human ingenuity more than **metals**. Used as currency, jewelry, and weaponry, civilizations power was correlated with use of the material (or the lack thereof). Not surprisingly, metals also make up over 80 percent of all the elements in existence. Indeed, the mastery of metals was once commonly used to designate the ages of civilization (e.g., Stone Age to Bronze Age to Iron Age), indicating advancement in toolmaking capabilities (a rough measure of engineering technology).

That being said, because of the lack of availability, prior to the beginning of the nineteenth century, metals had little *structural* role in buildings and large-scale structures, except in connections and nominal reinforcement. This is mainly due to the fact that the production of metals had to "catch up" to the societal demand. The Industrial Revolution was characterized by new manufacturing processes in the period from about 1760 to about 1840. This transition included going from hand production methods to machines, new chemical manufacturing and iron production processes, and the development of machine tools. The stage was set for metals to eclipse all other materials in terms of usage and applications.

Metallurgy and Materials Science

Society's technological needs have driven the history of material science—as construction, transportation, and industrialization expanded, procurement and production of metals were the primary focus. No substance has been as important as metal in the story of man's control of the built environment. Advances in agriculture, warfare, transport, and even cookery are impossible without metal. So is the entire Industrial Revolution, from steam to electricity. Other materials, such as ceramics and polymers, were only recently developed (the twentieth century), and many of their potential uses in applications were not seen until after World War II. It is not too broad a statement to say that the recent field of materials science and engineering grew out of the studies of **metallurgy**.

Metallurgy was traditionally a domain of materials science and materials engineering that studies the physical and chemical behavior of metallic elements, their intermetallic compounds, and their mixtures, which are called **alloys**.[1] The production of metallic materials involves the processing of ores to extract the metal they contain, along with the mixture of metals to produce alloys. Certain metals can be recovered from their ores by simply heating the rocks in a fire, such as tin, lead, and copper, a process known as smelting. Therefore, these were the earliest metals exploited. About 3500 BC, it was discovered that by combining copper and tin, a superior metal could be made, an alloy called bronze, representing a major technological shift, which began the Bronze Age.

More recently, metallurgy also encompasses the *technology* of metals: the way in which science is applied to the production of metals for use in products for consumers and manufacturers. Metals in general have high electrical conductivity and high thermal conductivity, and thus are typically used in electronic and technological applications. Typically, they are malleable and ductile, deforming under stress without cleaving, which facilitates making device parts or components. In terms of optical properties, metals are shiny and lustrous. These features are a result of the atomic bonding and crystalline structure of metals, as discussed in **chapter 3**. In fact, the particular atomistic structure is what defines metals as metals (those bonded by metallic bonds, of course). Of the 118 (current) elements on the periodic table, 91 of them are classified as metals. Some common metals and their uses are listed in **table 5.1**.

While metals are typically shiny and feature good electrical and thermal conductivity, we typically don't care about those properties in civil or structural engineering. More importantly, from a *mechanical* perspective, metals are generally as follows:

1. **Malleable**: They can be hammered or pressed permanently out of shape without breaking or cracking
2. **Fusible**: Able to be fused together (or melted to separate)
3. **Ductile**: Able to be drawn out into a thin wire

These properties allowed civilizations to engineer metals into building components and combine the components into larger (relatively safe and predictable) structures. Metals then became structurally important for many common systems and goods (e.g., cars, rebars

---

1 A mixture or solid solution composed of a metal and another element.

TABLE 5.1 COMMON METALS AND THEIR USES

Metal	Symbol	Properties/Uses/Features
Aluminum	Al	The lightness, strength, and corrosion resistance of aluminum are important considerations in its application. Metallic aluminum is used in transportation, packaging such as beverage cans, building construction, electrical applications, and other products.
Beryllium	Be	Beryllium is very light, has a high melting temperature, is an ideal metal for use in the aerospace and defense industries, and is almost always alloyed with other metals. Another application is in the manufacture of gasoline pumps, because an alloy of copper and beryllium does not spark when hit against other metals.
Chromium	Cr	Chromium is alloyed with steel to make it corrosion resistant or harder. An example is its use in the production of stainless steel, a bright, shiny steel that is strong and resistant to oxidation (rust). Stainless steel consumes most of the chromium produced annually.
Cobalt	Co	Cobalt has been used by civilizations for centuries to create beautiful deep-blue glass, ceramics, pottery, and tiles.
Copper	Cu	Pure copper is drawn into wires or cables for power transmission, building wiring, motor and transformer wiring, wiring in commercial and consumer electronics and equipment, and telecommunication cables. It is also used for plumbing, heating, and air conditioning tubing.
Gallium	Ga	Gallium is used in a variety of highly specialized electrical applications. Gallium and gallium alloys are used for lasers, photo detectors, light-emitting diodes, solar cells, and highly specialized integrated circuits, semi-conductors, and transistors.
Gold	Au	Most gold is used to make jewelry and other art items. Because it is chemically stable and conducts electricity so well, it is very important in electronics.
Iron	Fe	Almost all of the iron ore that is mined is used for making steel.
Lithium	Li	More than one-half of the lithium compounds consumed are used in the manufacture of glass, ceramics, and aluminum. Lithium is also used in making synthetic rubber, greases, and other lubricants. Lithium batteries are proving to be an effective and affordable alternative to traditional batteries, including in new battery applications.
Platinum	Pt	Most platinum is used to produce catalytic converters in automobile exhaust systems. Although about one-third of all platinum is used by the automotive industry, there are various other uses. It is alloyed with gold, silver, and copper for dental uses.
Titanium	Ti	Titanium is lighter than steel but still is very strong. It also has a very high melting temperature. These physical properties make titanium and titanium alloys very useful in the aerospace industry, where it is mostly used to make engines and structural components for airplanes, satellites, and spacecraft.
Tungsten	W	Tungsten is mixed with carbon to make a very strong, very resistant material called tungsten carbide. Tungsten carbide is used to make cutting tools and wear-resistant tools for metalworking, drilling for oil and gas, mining, and construction.
Zinc	Zn	Zinc is relatively nonreactive in air or water. Consequently, it is applied in thin layers to iron and steel products that need to be protected from rusting. This process is called galvanizing.

for buildings, window frames, kettles, refrigerators). Of all the metals used for structural applications, none are as widespread, well understood, and important as *steel*.

Steel is used widely in the construction of roads, railways, other infrastructure, appliances, and buildings. Most large modern structures, such as stadiums, skyscrapers, bridges, and airports, are almost universally supported by a steel skeleton. As will be discussed in **chapter 6**, even those with a concrete structure employ steel for reinforcing.

Due to its importance within civil engineering, this chapter concentrates on steel. That being said, with slight variations, the processing and mechanical behavior of other metals are similar to steel, and thus it serves as a platform for all metals in general. To design a steel structure, an engineer must be knowledgeable in the limitations of the material, requiring understanding of the material production, effects of treatment, mechanical behavior, and failure across the lifetime of the material.

II. Steel and Steel Production

When you compare iron and steel with something like aluminum, you can see why it was so important historically. To refine aluminum, you need access to huge quantities of electricity (energy), and it is difficult to cast and/or shape. Iron, however, is much easier to manipulate. The element has been useful to people for thousands of years, while aluminum really didn't exist in any meaningful way until the twentieth century. Iron can be created relatively easily with tools that were available to early societies. There may come a day when humans become so technologically advanced that iron is completely replaced by exotic metals and plastics. But right now, steel is king and has many advantages over these much more expensive alternatives.

As early as the eleventh century BC, it was known that with a little know-how, the properties of pure iron (Fe) can be much improved. Perhaps serendipitously, it was discovered that if iron is reheated in a furnace with charcoal (containing carbon, C), some of the carbon is transferred to the iron (Fe + C). This process subsequently **hardens** the metal, and the effect is considerably greater if the hot metal is rapidly reduced in temperature, usually achieved by **quenching** it in water.[2] The new material produced is **steel**. It can be worked (or "wrought") just like softer iron, and it will keep a finer edge, capable of being honed to sharpness.

Of all metals, steel is perhaps the first that comes to mind when you think of engineering materials. Throughout the 20th century, steel became an increasingly important engineering

---

2   Actually, *saltwater* or *brine* was a better medium to quench the hot steel because freshwater would form a layer of steam at the surface that would reduce the cooling effect. This led to many different fluids being used as quenching baths throughout the ages, including horse urine and blood.

material because it is **precise**, **machinable**, and, above all, **predictable**, as well as light in proportion to its weight. Therefore, steel is well suited as a building block for engineered structures. Cars, tractors, bridges, trains (and their rails), tools, skyscrapers, weapons, and ships all depend on iron and steel to make them strong. Particularly for civil engineering, steel is the metal of most interest and importance.

Steel has gained this level of popularity for a number of reasons, including the following:

1. The amount of available ore (in which steel is refined) and the relatively low-cost refinement process meant that steel was one of the first metals easily manufactured in large quantities.
2. The mechanical properties and behavior of steel are highly conducive to structural engineering, unmatched by any other material in terms of cost/performance.
3. The availability of preformed structural steel members including beams, columns, struts and joists, ties, hangers, and cables, as well as the established methods of connections (e.g., bolts, welds) make steel easy to work with in a wide variety of applications.

Beyond purely steel structures, we note that steel is the primary reinforcement in concrete because of concrete's relatively low tensile strength (as will be discussed in **chapter 6**).

Steel Production and Refinement

Steel, by definition, is an iron and carbon alloy that typically contains *less than* 2 percent carbon (by weight). While this seems like an extremely nominal amount, we can consider the atomic weights of both iron (Fe) and carbon (C) and see otherwise. The atomic weight of iron atoms is approximately 56 amu, while the atomic weight of carbon atoms is approximately 12 amu. Thus, 2 percent carbon by weight is approximately one carbon atom per ten irons, or 10 percent by concentration. In terms of added atoms, this is a significant amount in the crystal lattice of iron. If we consider other impurities in the steel (such as traces of phosphorous, sulfur, oxygen, and nitrogen), we clearly see that the metal we call "steel" is not a uniform and homogeneous material such as "gold" or "silver." Indeed, classification of steel type is based on the quantity of carbon and other additives.

Plentiful, consistent, and inexpensive steel was made possible by the **Bessemer process**, developed in the mid-nineteenth century, which used air injected into molten iron to remove impurities (one cause of the brittleness of cast iron).

To produce steel, iron is first required, which is attained from **iron ore** (any rocks and minerals from which metallic iron can be extracted). Iron ore is mined (typically iron oxides, sulfides, carbonates, silicates) in large open pits that uncover iron-rich banded deposits in the ground. Currently, over 95 percent of the globally mined iron ore is used to make steel. The most common ores contain the iron oxides hematite (Fe_2O_3) and magnetite (Fe_3O_4).

Iron ore is then mixed with coal (predominantly carbon) and limestone and put in a blast furnace.[3] The furnace gets its name from the method that is used to heat it. Preheated air at about 1,000°C is blasted into the furnace through nozzles near its base. The coke is burned and produces carbon monoxide, which reacts with the iron oxides to reduce the elemental iron (Fe), or

$$Fe_2O_3 + 3CO \rightarrow 2Fe + 3CO_2.$$

Because the furnace temperature is extremely high, the metal is produced in a molten state, and this runs down to the base of the furnace. Of note, the temperature required in this process is on the order of 1,500°C to 1,700°C; such temperatures are attainable via charcoal fires.[4] Metals with extremely high melting temperatures, such as tungsten, would not be a useful structural material.

Limestone ($CaCO_3$) is used as the classical method to carry impurities away (limestone plus impurities is called the **slag**), but is not efficient, leaving many impurities behind. This process typically yields iron with high carbon content (e.g., pig iron). Pig iron contains 4 to 5 percent carbon and is extremely hard and brittle. One historical option was to melt it, mix it further with slag, and hammer it out to eliminate most of the carbon to create strong, malleable **wrought iron**. Another process was to melt the pig iron and combine it with scrap iron, smelt out impurities, and add alloys to form **cast iron** (2 to 4 percent carbon). Cast iron and wrought iron were used increasingly for framing industrial buildings in the early nineteenth century, but the use was limited due to the unpredictable brittleness of cast iron and the relatively high cost of wrought iron.

To attain steel, it is necessary to reduce the levels of carbon and other impurity elements in the hot metal. Molten steel is most commonly refined in ladles in the foundry. Some of the operations performed in ladles include deoxidation, vacuum degassing, alloy addition, inclusion removal, inclusion chemistry modification, desulphurization, and homogenization. Tight control of ladle metallurgy, temperature, and chemistry is associated with producing high grades of steel in which the tolerances in chemistry and consistency (e.g., carbon content and alloys) are narrow.

---

3   Note that the process described here is considered *basic oxygen steelmaking*, also known as Linz-Donawitz-Verfahren steelmaking, which is a variation of the historical Bessemer process. There are other steelmaking methods currently being used; however, the basic oxygen method is the most common.

4   Charcoal burns at temperatures up to 2,700°C. Because of its porosity, it is sensitive to the flow of air, and the heat generated can be moderated by controlling the airflow to the fire. For this reason, charcoal is an ideal fuel for a forge, for the production of iron and has been used that way since Roman times.

Again, by definition, steel has carbon content, and this content greatly affects the mechanical properties of the steel (even small changes—properties are very sensitive to changes in carbon). As a general rule, as the carbon content increases, the strength and the hardness of the metal increases (↑), while the ductility decreases (↓). The carbons, while assuming a nice atomistic "fit" in the crystal structure of iron, act as effective deformation and dislocation barriers when subject to stress. Thus, an increase in strength is associated with more carbons (more barriers) but a more brittle failure. Too little carbon, and the metal is soft and weak. Wrought iron, for example, has high workability both because of low carbon content (less than 0.1 percent) and slag mixed in (3 percent). A summary of steels and associated carbon contents is given in **table 5.2**.

Mild steel, also called plain-carbon steel, has carbon content on the order of 0.05 to 0.25 percent, making it malleable and relatively ductile. Mild steel is the most common form of steel because its price is relatively low, while it provides material properties that are acceptable for many applications. While mild steel has a relatively low tensile strength, it is cheap and malleable and is thus often used when large quantities of steel are needed (e.g., structural steel).

Additional carbon necessitates a consideration of strength, ductility, durability, and cost, as adding the carbon content requires additional processing. **Medium carbon steel** (0.3 to 0.6 percent), for example, is typically used for automotive components, while **high-carbon steels** (0.6 to 2.0 percent) are very strong and used for springs, high-strength wires, and machine tools that require great strength and hardness.

Iron-Carbon Phase Diagram

The percentage of carbon present and the temperature define the phase of the iron-carbon alloy and therefore its physical characteristics and mechanical properties. The percentage of carbon determines the type of ferrous alloy: iron, steel, or cast iron. As seen in **chapter 4**, a **phase diagram** is a type of chart used to show conditions at which thermodynamically distinct phases of an alloy can occur at equilibrium. While the intent here is not to describe

TABLE 5.2 GENERAL BREAKDOWN BASED ON CARBON CONTENT

Type	Carbon Content (percent weight)
Wrought iron	0–0.1
Mild steel	< 0.25
High-carbon steel	0.6–2.0
Pig iron	3.5–4.5
Cast iron	5.0

the details of the metallic phase transformations, a basic understanding of the mechanisms, time, and temperature dependencies, as well as the consequences for the most common heat treatments is essential to better appreciate the performance of steel in engineering applications.

As steel is an alloy of iron (Fe) and carbon (C), a study of the constitution and structure of all steels must first start with the iron-carbon (Fe-C) equilibrium diagram (**figure 5.1**). Many of the basic features of this system influence the behavior of even the most complex alloy steels, as well as a number of various heat treatments they are usually subjected to (e.g., hardening, annealing). The iron-carbon diagram provides a valuable foundation on which to build knowledge of both plain-carbon and alloy steels in their immense variety. Since the phase diagram is limited in the content of carbon (e.g., it does not span to 100 percent carbon content), it is effectively an iron-iron carbide (Fe-Fe$_3$C) phase diagram.

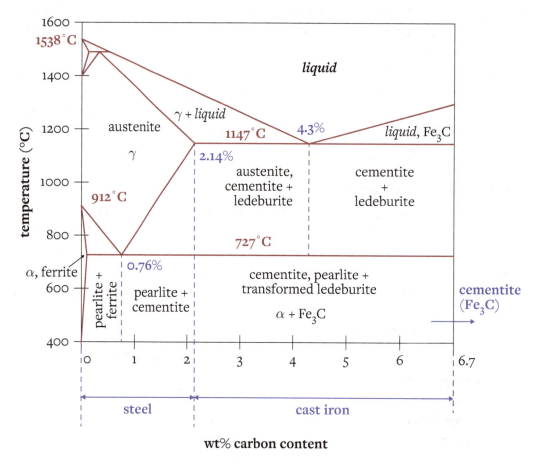

FIGURE 5.1 Example Fe-C phase diagram.

Unlike the previous phase diagrams considered, it is noted that the equilibrium diagram effectively represents a metastable equilibrium between iron and iron carbide (cementite). Cementite is metastable, and the true equilibrium should be between iron and graphite (e.g., 100 percent carbon). Although graphite occurs extensively in cast irons (2 to 4 percent weight C), it is usually difficult to obtain this equilibrium phase in steels with lower carbon contents (0.03–1.5 percent weight C). The much larger phase field of γ-iron (austenite) compared with that of α-iron (ferrite) reflects the much greater solubility of carbon in γ-iron. The α-iron phase and δ-phase are severely restricted, with limited carbon solubility and temperature ranges. Some key phases include cementite, ferrite, austenite, and pearlite, which are described next:

- **Cementite**, also known as iron carbide, is a chemical compound of iron and carbon, with the formula Fe_3C (or Fe_2C:Fe). By weight, it is 6.67 percent carbon and 93.3 percent iron. It has an orthorhombic crystal structure. It is a hard, brittle material, normally classified as a ceramic in its pure form, though it is more important in metallurgy. In carbon steel, it either forms from austenite during cooling or from martensite during tempering.
- **Ferrite**, also known as α-ferrite (α-Fe) or α-iron, is a materials science term for pure iron, which is a ductile material with a body-centered cubic (BCC) crystal structure. It is this crystalline structure that gives steel and cast iron their magnetic properties and is the classic example of a ferromagnetic material.
- **Austenite**, also known as gamma phase iron (γ-Fe), exists above a critical temperature of approximately 1,000 K (1,340°F; 730°C). From 912 to 1,394°C (1,674°F to 2,541°F), α-iron undergoes a phase transition from BCC to the face-centered cubic (FCC) configuration of γ-iron. This is similarly soft and ductile but can dissolve considerably more carbon. This gamma form of iron is exhibited by the most commonly used type of stainless steel for making hospital and food service equipment.
- **Pearlite** is a two-phased lamellar (or layered) structure composed of alternating microscopic layers of α-ferrite (approx. 88 percent weight) and cementite (approx. 12 percent weight) that occurs in some steels and cast irons. Pearlite is a common microstructure occurring in many grades of steels. Steels with pearlitic or near-pearlitic microstructure can be drawn into thin wires. Such wires, often bundled into ropes, are commercially used as piano wires, cables for suspension bridges, and steel cord for tire reinforcement. High degrees of wire drawing (a true strain above 300 percent) leads to pearlitic wires with yield strengths of several gigapascals. It makes pearlite one of the strongest structural bulk materials on earth.

Heat Treatments

Various **heat treatments** are used to alter the properties and composition of steel. Very few metals react to heat treatment in the same manner, or to the same extent, that carbon steel does, and steel heat treating behavior can vary radically depending on alloying elements. As such, the descriptions given here are only meant to be representative of complex metallurgic processes.

Steel can be softened to a very malleable state through annealing, or it can be hardened to a state nearly as rigid and brittle as glass by hardening or quenching. However, in its hardened state, steel is usually far too brittle, lacking the structural integrity to be useful for most applications. Tempering is a method used to decrease the hardness, thereby increasing the ductility of the quenched steel to impart some "springiness" and malleability to the metal. This allows the metal to bend before breaking. Depending on how much temper is imparted to the steel, it may bend elastically (the steel returns to its original shape once the load is removed), or it may bend plastically (the steel does not return to its original shape, resulting in permanent deformation) before fracturing. Tempering is used to precisely balance the mechanical properties of the metal, such as shear strength, yield strength, hardness, ductility, and tensile strength, to achieve any number of a combination of properties, making the steel useful for a wide variety of applications.

The term **annealing** refers to a heat treatment in which a material is exposed to an elevated temperature for an extended period of time and then cooled. Typically, the target temperature is sufficient to induce a phase transformation (which differs depending on the composition of the steel).

Normalizing is a type of annealing treatment, with the difference based on temperature used. By holding the temperature significantly above the transition temperature (over 100 K), crystal grains can reorganize. Normalized heat treatment establishes a more uniform, fine-grained steel, which facilitates later heat treatment operations and produces a more uniform final product. Smaller grains form that produce a tougher, more ductile material (used in structural plate production for high-fracture toughness).

Full annealing heat treatment differs from normalizing heat treatment in that the temperature is typically only 10 K to 40 K above the critical transition temperature, and the cooling rate is slower. The process consists of heating the steel to the austenite-stable range, holding, and then slowly cooling. This establishes a soft microstructure and thus a soft product. The goals of annealing are to refine the grain and soften the steel, and to remove internal stresses and gases, which ultimately increases ductility and toughness.

Hardening (or specifically for steel, **quench hardening** or **quenching**) is a process that increases the hardness of a metal. The typical method for steel is to heat up high-carbon content steel such that the carbon and iron form a new phase, and then "quench" (i.e., rapidly cool) the metal. It is important to quench with a high cooling rate so that the carbon does not have time to form precipitates of carbides. All hardening mechanisms introduce crystal lattice defects that act as barriers to dislocation slip. While the quenching process hardens steel, it can cause high internal strains; it is commonly followed by tempering to alleviate internal stresses.

Tempering is another process of heat treating, usually performed after hardening, to reduce some of the excess hardness imbued by internal stresses caused by distorted crystal structures. Tempering is accomplished by heating the metal to a much lower temperature than was used for hardening and effectively increases the ductility and toughness of the steel. The exact temperature determines the amount of hardness removed and depends on both the specific composition of the alloy and the desired properties in the finished product. For instance, very hard tools are often tempered at low temperatures, while springs are tempered to much higher temperatures.

III. Steel Designations

As we have seen, the term "steel" refers to a family of materials, not just a single element such as silver or gold. In engineering practice, stating "the beam is made of steel" is only a partial description. More likely, such statements will be stated as "the beam is A36 S18 × 70." To make sense of that, we need to discuss the standard designations of steel.

Steel Alloys

Since steel is, by definition, composed of iron and carbon, every steel is truly an alloy, but not all steels are called "alloy steels." Even the simplest steels are iron (Fe) (about 99 percent) alloyed with carbon (C) (about 0.1 to 1 percent, depending on type). However, the term "alloy steel" is the standard term referring to steels with other alloying elements *in addition* to the carbon. Common alloyants include manganese (the most common one), nickel, chromium, molybdenum, vanadium, silicon, and boron. Less common alloyants include aluminum, cobalt, copper, cerium, niobium, titanium, tungsten, tin, zinc, lead, and zirconium. These elements are typically added at the end of steel processing.

The following is a range of improved properties in alloy steels (as compared to carbon steels): strength, hardness, ductility, toughness, wear resistance, machineability, corrosion resistance, and hardenability. Some common alloying agents and their effect are described in **table 5.3**.

To achieve some of these improved properties, the metal may require heat treating. Low-alloy steels (less than 7 percent) are usually used to achieve better hardenability,

TABLE 5.3 COMMON ALLOYING AGENTS AND MECHANICAL EFFECT

Element	Effect
Manganese (Mn)	Resistance to abrasion and impact
Molybdenum (Mo)	Strength
Vanadium (V)	Strength and toughness
Nickel (Ni)	Toughness and stiffness
Chromium (Cr)	Toughness and stiffness

which in turn improves its other mechanical properties. They are also used to increase corrosion resistance in certain environmental conditions. With medium- to high-carbon levels, low-alloy steel is difficult to weld. Lowering the carbon content to the range of 0.10 to 0.30 percent, along with some reduction in alloying elements, increases the weldability and formability of the steel while maintaining its strength. Such a metal is classed as a high-strength, low-alloy steel. Some of these find uses in exotic and highly demanding applications, such as in the turbine blades of jet engines, in spacecraft, and in nuclear reactors. Because of the ferromagnetic properties of iron, some steel alloys find important applications where their responses to magnetism are very important, including in electric motors and in transformers.

Common Alloy: Stainless Steel

Stainless steel is a material that we are familiar with and come across in a vast array of everyday items—from cutlery, jewelry, and razor blades to washing machines and cars. The term "stainless" was coined early in the development of the material for cutlery applications that did not corrode or lose their luster; the main requirement for stainless steels is that they should be *corrosion resistant* for a specified application or environment. The corrosion resistance and low porosity of stainless steel also has the added benefit of being easy to sterilize; as such, it is frequently used in medical instruments (**figure 5.2**) and medicine containers, as it will not corrode and cause contamination of the contents. For similar reasons, stainless steel is also used inside the human body in implants such as hip joints.

The technical requirement for a steel to be classed as stainless is a minimum weight of 10.5 percent chromium (Cr). This can be more than doubled for harsh environments, and other alloying elements such as nickel (Ni) are sometimes added to enhance its structure, and properties (see **table 5.3**). Stainless steel must be used in an oxygenated environment, and it is the chromium that oxidizes to form a layer of Cr_2O_3. This layer has very different attributes than the iron-oxide film that forms on unprotected carbon steels that we commonly

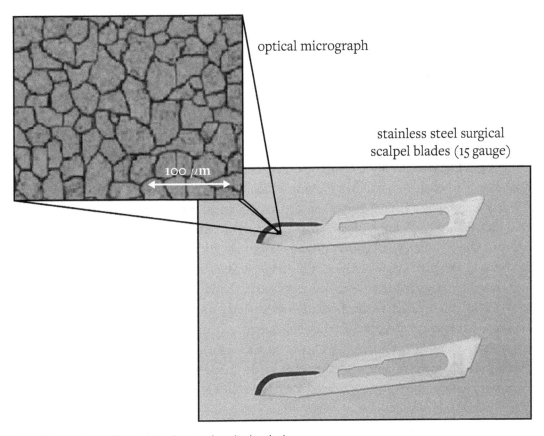

optical micrograph

stainless steel surgical
scalpel blades (15 gauge)

100 μm

FIGURE 5.2   A 45-millimeter stainless steel medical scalpel.

call rust. First, the layer of chromium oxide is too thin to be seen. Second, there is sufficient chromium to form a passive film of oxide, which prevents further surface corrosion and its spread into the bulk material.

Being 100 percent recyclable, stainless steel has become a big player in sustainability. About 60 percent of stainless steel used in products today is recycled—some from items that have reached their end of life, while some is scrap from manufacturing processes used to make other stainless steel products.

Structural Steel

While steel can be formed in uniform plates and bars, more refined shapes are typically implemented for structural purposes. Structural steel components are produced in a structural mill, where hot steel blanks pass through a succession of rollers that gradually press the metal into progressively more refined approximations of the desired cross-section. The machine/roller spacings in the mill can be adjusted, resulting in a range of possible member cross-sectional dimensions. This range provides the structural engineer with a finely graduated selection of shapes from which to select an appropriate component with the desired capacity and dimension.

Structural Shapes

Different structural shapes are used in structural engineering, depending on the functional role of the member (e.g., a beam section would have different requirements than a truss or anchorage). Some of the more common examples are given in **table 5.4.**

Due to their ease of production (symmetrically rolled) and efficient cross-section in flexure/bending, wide-flange type I-beams are the most common shapes for beams and columns. Wide flanges are available in a vast range of sizes (depths of 4 to 40 inches) and weights (approximately 10 to 750 pounds per linear foot). If needed, welding together flanges and web plates may produce even larger, heavier sections.

Steel angles (L-shaped) are smaller and weaker than I-sections, but are extremely versatile. They can be used as short beams supporting small loads, or as reinforcement for concrete or masonry structures. They also serve as connection or bracing elements for larger W-sections (e.g., reinforcing a beam-column joint). Angles can also be built up into trusses (commonly, two angle sections are welded back-to-back such that they maintain symmetry and can be easily connected to gusset plates at the joints of the truss).

Beyond such cross-sections, other common structural steel components include sheet piling sections and rails for trains, and they have their own specialty designations.

Structural Steel Grades

Structural steels have a detailed classification system. Mild structural steel, known by its ASTM designation as A36, is the most predominant type used in steel building

TABLE 5.4   COMMON STRUCTURAL STEEL SHAPES AND DESIGNATIONS

Shape	Symbol	Designation	Explanation
Wide flange (I-beam)	W, HP, M	W 21 × 83	21: nominal depth in inches; 83: weight per foot of length in pounds (lb/ft)
American standard (I-beam)	S	S 18 × 70	18: nominal depth in inches; 70: weight per foot of length in pounds (lb/ft)
Angle	L	L 4 × 3 × 3/8	4 × 3: lengths of the two legs; three-eighths thickness
Channel	C, MC	C 9 × 13.4	9: nominal depth in inches; 13.4: weight per foot of length in pounds (lb/ft)
Structural tee	WT, ST	WT 13.5 × 47	Tee cross-section by splitting a W 27 × 94

frames. With the addition of other elements similar to stainless steel, high-strength, low-alloy structural steels can be produced. **Table 5.5** lists some common steel grades and associated properties.

As seen in **table 5.5**, the most common benefit of alloying structural steel is increasing both yield strength and corrosion resistance. However, the elastic modulus, E, is typically not affected by added elements (where $E = 200$ GPa for all steel grades in **table 5.5**). This is due to the chemical structure of steel—the "pinning" mechanisms of alloying agents only act after plastic yield. The bond strength (stiffness) does not vary much. This is critical if serviceability requirements are the controlling factor of design; if you do not need the extra strength, then a higher-grade steel is not cost effective.

Reinforcing Steel

Reinforcing steel—as the name implies—has no structural function independently, but rather serves as reinforcement for concrete and masonry structures. While concrete, for example, is strong in compression, it is relatively week in tension (as discussed in the next chapter). To make a strong flexural member (which is subject to both tensile and compressive stresses), steel reinforcing bars, or **rebars**, are placed along the axis of bending to resist the tensile stresses. In effect, this makes all concrete structures and elements composite materials, and they are commonly referred to as reinforced concrete, or simple "RC" structures.

TABLE 5.5   COMMON STEEL GRADES AND PROPERTIES

ASTM Grade/Alloy	Description	Yield Strength σ_y
A36	Mild, structural grade carbon steel	36,000 psi (248 MPa)
A242	High strength, low alloy, *corrosion resistant*	42,000–50,000 psi (290–345 MPa)
A514*	High-yield strength quenched and tempered	90,000–100,000 psi (620–690 MPa)
A529	Structural carbon steel	min. 42,000 psi (min. 290 MPa)
A572	High strength, low alloy	42,000–65,000 psi (290–448 MPa)
A588	High strength, low alloy, *corrosion resistant*	min. 50,000 psi (min. 345 MPa)

*Only available in plates

One of the odd facts of reinforcing steel is that it is completely invisible. You never actually see it because it's always buried/embedded, yet it is frequently used in many structures and a key element for strength. Concrete reinforcing steel is used in bridges, buildings, skyscrapers, homes, warehouses, foundations, and roads to increase the strength of the concrete and ultimately help hold up the structures. While concrete alone is strong, reinforcing steel significantly increases the strength of concrete in an economical and safe manner. Steel has a thermal expansion coefficient nearly equal to that of modern concrete. If this were not so, it would cause problems through additional longitudinal and perpendicular stresses at temperatures different from the temperature of the setting. Reinforced concrete is also considered environmentally friendly, as the steel is (commonly) made from 100 percent recycled scrap. At demolition, the reinforcing steel and the concrete are separated and recycled for reuse.

Steel reinforcing bars are produced by pouring molten steel into casters and then running it through a series of stands in the mill, which shape the steel into reinforcing bars. Of note, standard rebars are typically deformed—they have patterned surface protrusions, or cross-hatchings, which are called "deformations." The deformations help secure the steel and transfer load between it and the concrete (i.e., the deformations facilitate a better bond between the steel and concrete). Deformed bars are used for beams, columns, slabs, walls, footing, bridge decks, and brickwork.

Types and Designation of Reinforcement Steel

Most steel reinforcement is conceptually divided into primary and secondary reinforcement, but there are other minor uses:

- **Primary reinforcement** refers to the steel that is employed specifically to guarantee the necessary resistance needed by the structure as a whole to support the design loads.
- **Secondary reinforcement**, also known as distribution or thermal reinforcement, is employed for durability and aesthetic reasons by providing enough localized resistance to limit cracking and resist stresses caused by effects such as temperature changes and shrinkage.
- **Secondary reinforcement** is also employed to confer resistance to concentrated loads by providing enough localized resistance and stiffness for a load to spread through a wider area.

In addition, both primary and secondary reinforcing reduces random cracking, reduces and controls crack width, and helps maintain aggregate interlock. Strength is increased with steel-reinforced concrete—even the smallest cross-sectional area of steel reinforcement will provide reserve strength of 16 percent and more.

Common rebar is made of unfinished tempered steel, making it susceptible to rusting. As rust takes up greater volume than the steel from which it was formed, it causes severe internal pressure on the surrounding concrete, leading to cracking, spalling, and, ultimately, structural failure. This is a particular problem where the concrete is exposed to saltwater, as in bridges where salt is applied to roadways in winter, or in marine applications. Uncoated, corrosion-resistant, low-carbon/chromium (micro-composite), epoxy-coated, galvanized, or stainless steel rebars may be employed in these situations at greater initial expense, but at significantly lower expense over the service life of the project.

In the United States, rebar sizes are given a number related to the diameter of the bar. Imperial bar sizes give the diameter in units of one-eighth an inch, so that #8 = 8(1/8 inch) = 1-inch diameter. The cross-sectional area, as given by πr^2, works out to (bar size/9.027)^2, which is conveniently approximated as (bar size/9)^2 square inches. For example, the area of #8 bar is $(8/9)^2 = 0.79$ sq in. Common rebar designations are given in **table 5.6**.

Like structural steel, rebar is available in different grades and specifications that vary in yield strength, ultimate tensile strength, chemical composition, and percentage of elongation. The grade designation is typically equal to the minimum yield strength of the bar in ksi (1,000 psi). For example, grade 60 rebar has a minimum yield strength of 60 ksi. Rebar is typically manufactured in grades 40, 60, and 75.

TABLE 5.6 US REBAR SIZE CHART

Imperial Bar Size	"Soft" Metric Size	Mass per Unit Length (lb/ft)	(kg/m)	Nominal Diameter (in)	(mm)	Nominal Area (in²)	(mm²)
#2	#6	0.167	0.249	0.250	6.35	0.05	32
#3	#10	0.376	0.561	0.375	9.525	0.11	71
#4	#13	0.668	0.996	0.500	12.7	0.20	129
#5	#16	1.043	1.556	0.625	15.875	0.31	200
#6	#19	1.502	2.24	0.750	19.05	0.44	284
#7	#22	2.044	3.049	0.875	22.225	0.60	387
#8	#25	2.670	3.982	1.000	25.4	0.79	509
#9	#29	3.400	5.071	1.128	28.65	1.00	645
#10	#32	4.303	6.418	1.270	32.26	1.27	819

Alternatives to standard "rebars" include plain and deformed wire fabrics (welded wire fabrics), slabs, and wires or cables, depending on application and/or design requirements.

Common Reinforcement Applications

Rebar is typically used as reinforcement in three distinct ways (conventional, prestressed, or post-stressed). However, regardless of application, the function of the reinforcement is to *decrease* the amount of tensile stress in the concrete component.

- **Conventional reinforcement** is steel rebar that is spaced and laid in concrete formwork and tied together, then concrete is typically poured around it, filling spaces and gaps. The rebar is stress-free when it is laid and typically placed where the concrete cross-section will experience tension when subject to design loads. For example, for a simply supported beam, the main reinforcement is placed at the bottom of the beam. The amount of tension in the steel is a function of the load on the concrete members. When the concrete is unloaded, the steel is also unloaded.

- **Prestressed reinforcement** is designed and placed such that the steel is intentionally tensioned prior to concrete curing. Usually, the steel is affixed to large anchor blocks and tightened accordingly. Concrete is cast around the reinforcement, and then when the concrete cures, the steel is released from the anchors. This typically causes compression in the section of the concrete that is reinforced (i.e., a beam will camber upward). When the concrete member is subject to load, the stress must first overcome this compressive stress before the concrete is subject to tension. A prestressed system is more difficult to design because the stresses in the steel and concrete are not a function of the applied loads (i.e., when the concrete is unloaded, there are still stresses). Prestressed reinforcement includes wires, strands, cables, and bars (with low relaxation). Advantages of prestressed concrete include crack control and lower construction costs. Costs are lower as designs can be satisfied with thinner slabs—especially important in high-rise buildings in which floor thickness savings can translate into additional floors for the same (or lower) cost and fewer joints, since the distance that can be spanned by post-tensioned slabs exceeds that of reinforced constructions with the same thickness. Increasing span lengths increases the usable floor space in buildings; diminishing the number of joints leads to lower maintenance costs over the design life of a building, since joints are the major focus of weakness in concrete buildings.

- **Post-stressed** or **post-tensioned reinforcement** follows a similar mechanism as prestressed reinforcement, except the tension can be adjusted after the concrete cures. Typically, the reinforcing steel (wires, cables, or rods) is set in independent channels/ducts within the concrete (i.e., is unbonded). Once the concrete has hardened, the tendons are tensioned by hydraulic jacks that react against the concrete member itself. One of the advantages of post-tensioning is the ability to individually adjust cables based on field conditions and greater geometric control of the placement (because of lack of tension when being laid). Post-tensioned systems require greater maintenance compared to conventional or prestressed reinforcements.

Steel Corrosion

One of steels' weaknesses is a tendency to corrode in certain environments; this decreases the stiffness of steel (causing unwanted deflections), weakens the strength of steel (which means it may not be able to carry the intended design loads), and may cause microcracking (which leads to fracture, to be discussed in the next section).

The corrosion of structural steel is an electrochemical process that requires the simultaneous presence of moisture (water) and oxygen. Essentially, the iron in the steel is oxidized to produce rust, which occupies approximately six times the volume of the original material. The rate at which the corrosion process progresses depends on a number of factors, but principally the "microclimate" immediately surrounding the structure. While more complex, the sum of the corrosion process can be expressed as follows:

$$4Fe + 3O_2 + 2H_2O \rightarrow 2Fe_2O_3H_2O$$

$$(iron) + (oxygen) + (water) \rightarrow (rust).$$

Water is the electrolyte needed for this reaction process, but salt (ions) enhances this. As such, corrosion is critical in steel structures close to saltwater (e.g., coastal areas). Moreover, contaminants in the air contribute to corrosion, including acid rain, coal burning, and chemical plants nearby.

Of note is that rusting itself provides a form of corrosion resistance: once the oxide forms, it partially protects the steel from further corrosion.

Methods for corrosion protection include the following:

- Barrier coatings: Paints, other surface treatments
- Inhibitive primer coatings: More advanced surface primers
- Sacrificial primers or cathodic protection: Most common in structural steel; an outer metal-based coating (e.g., zinc) corrodes instead of the steel (the reason structural steel already looks "rusted")

IV. Fracture

One of the critical issues with any structural design is that no material is perfect—they all have defects. While engineering calculations on paper can be arbitrarily precise, the material that gets fabricated and delivered on-site has inherent limitations in performance because of these defects. The worst-case scenario is **failure**, and, for metallic materials, one of the most common types of failure is due to **fracture** (**figure 5.3**). Without understanding fracture, stress analysis and design of metals (especially steels) would be of little use.

It is well known that the strength of metals (such as steel) can be increased by decreasing the grain size (i.e., the so-called Hall-Petch relation), which effectively inhibits dislocation motion (which was introduced in **chapter 4**). Unfortunately, this also results in a very **brittle material**—cracks can form at the grain boundaries and propagate catastrophically. This kind of brittle fracture can initiate at any kind of "stress concentrator," be it a grain boundary or other "crack," which can arise because of (a) material defects, (b) improper design, or (c) progressive failure.

Regardless of cause, fracture is typically initiated by an *overstress at a single point*, leading to *catastrophic crack extension*. It is this general case that we will consider for our introduction to fracture.

Physical Meaning of Fracture

Here we are considering the fracture of metals and ask, "What causes fracture?" First, we consider what happens when we apply stress to or deform a material: we increase the elastic

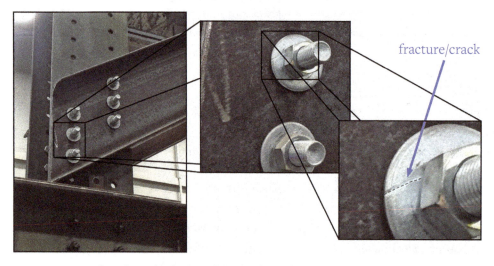

fracture/crack

FIGURE 5.3   Small cracks can lead to huge failures owing to fracture.

strain energy of the material. This stored energy is related to the external work done to the material (although, we are not concerned at this point with how the material is loaded). Second, we realize that, if there is a crack in the material, then the stress/strains cannot be transferred across the crack. If they could, what carries the load? Since there is no material there (by definition of a crack), the load must be carried around the crack itself (see **figure 5.4**). This typically leads to stress concentrations at the crack tips, as the "load" doesn't like to "travel far."

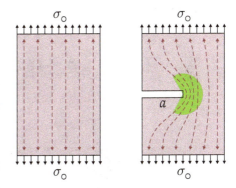

FIGURE 5.4   Stress path change because of a "crack."

Finally, from **chapter 4**, we know most metals have a crystalline form. For an FCC metal, for example, an atom is the most stable when it is bonded with the nearest neighbors (12 for FCC). Any deviation from this number of bonds results in an increase in energy (remember, bonds decrease the energy between two atoms). Thus, surface atoms have higher energy than atoms in the "bulk" because they have fewer bonds. This also means it takes a certain amount of energy to create a free surface. When a crack occurs in a metal, it is creating two free surfaces.

We can now start putting everything together and see why fracture occurs:

1. There is an energy increase owing to applied stress/deformation (elastic strain energy).
2. There is a localization of the elastic energy at a crack tip (because of the geometry).
3. There is an energy barrier to create a free surface in a crystal (i.e., extending a crack).

If the local strain energy owing to the stress and existing crack exceeds the barrier to create a free surface, then the crack extends. Typically, this has a cascading effect, as if the stress isn't released, the new (larger) crack localizes even more strain energy, and the crack extends again, and so on and so forth, leading to **fracture**. Thus, we can say when a crack reaches a certain critical length, it can extend almost instantaneously[5] through the structure, even at stresses less than the yield stress or ultimate tensile stress (UTS).

We wish to be able to use the energy balance of applied stress and surface creation to derive the critical fracture stress for a given crack/geometry. First, we will approximate the approach using a common civil engineering structure—**cantilever beams**.

---

5   The speed at which a crack propagates can actually be theoretically calculated, and for steel, it is on the order of 5,000 meters per second. The details behind this calculation are beyond the scope of this course.

Take-Out Chopsticks

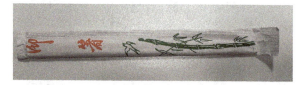

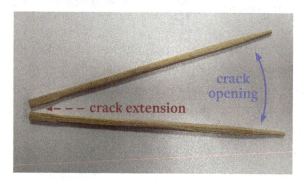

crack opening

crack extension

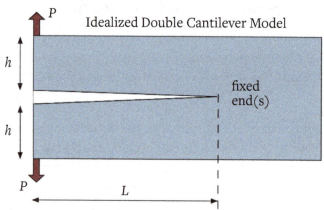

P

Idealized Double Cantilever Model

h

h

fixed
end(s)

P

L

FIGURE 5.5 Geometry of the "chopstick" fracture problem. We simplify the system to a pair of cantilever beams, with a length "*L*," height "*h*," and width "*b*" (into the page), fixed at a common point at their base. A load "*P*" is applied to both "beams" to separate them by fracturing the connection.

Cantilever Beam Derivation—"Splitting Chopsticks"

Here, we use a convenient (and solvable) engineering problem to demonstrate the onset of fracture. We consider disposable wooden chopsticks—the kind that are joined at the base and must be split apart prior to use (see **figure 5.5**). For analysis, we will model this system as two cantilever beams, joined at their fixed ends, with a load "*P*" applied to each beam in the opposite direction.

Here the "crack" is the free length of the cantilevers. We can derive the elastic energy because of the deflection of two cantilever beams[6] of length *L* as

$$U(L) = 2\left(\frac{P^2L^3}{3EI}\right).$$

Now we ask ourselves, "What happens if there is a little crack extension?" That is, the "crack" grows from length L to $L + \delta L$, as shown in **figure 5.6**. We know two things happen that affect the total energy (sometimes called the free energy) of the system.

---

6   How to calculate this is taught in structural analysis classes. It is equal to two times the work of the external load, *P*, multiplied by the deflection at the end of a beam, Δ, where $\Delta = PL^3/3EI$, such that $W = P\Delta$.

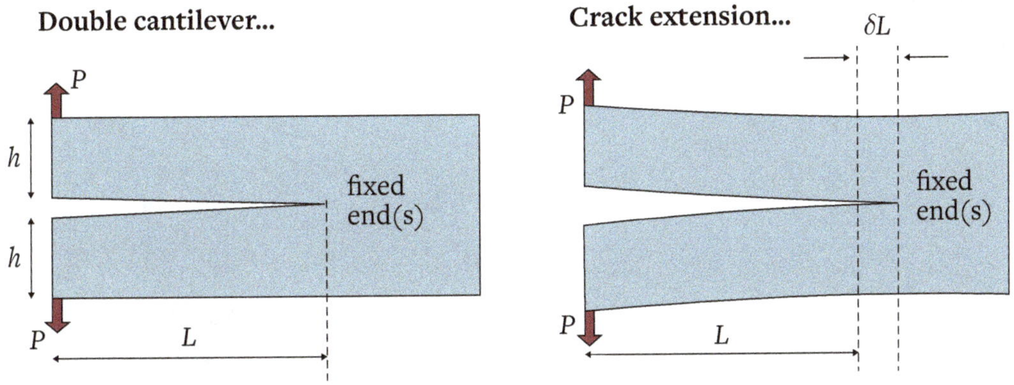

FIGURE 5.6 Extended crack geometry of the "chopstick" fracture problem. As the loads bend and separate the beams, the crack extends by δL, with a new total crack size of $L + \delta L$ and two new exposed material surfaces, with areas of $A = \delta L b$.

First, there are two extra surfaces created, requiring an increase in surface energy, δL:

$$\delta S = (\text{surface area})(\text{surface energy}) = 2\delta L b \gamma ,$$

where γ is the surface energy (e.g., J/m^2), δL the length of the new surfaces, and b the width of the new surfaces.

Second, there is an increase in elastic strain energy:

$$U(L + \delta L) = 2\frac{P^2(L + \delta L)^3}{3EI} ,$$

$$U(L + \delta L) = 2\frac{P^2\left(L^3 + 3L^2\delta L + 3L(\delta L)^2 + (\delta L)^3\right)}{3EI} ,$$

where δL is very small, $(\delta L)^2$ and $(\delta L)^3$ are approximately zero. Thus, the equation may be simplified to

$$U(L + \delta L) = 2\frac{P^2 L^3}{3EI} + 2\frac{P^2 L^2 \delta L}{EI} .$$

We want, however, the *change* in elastic energy, or

$$\delta U = U(L + \delta L) - U(L) = 2\left(\frac{P^2 L^2 \delta L}{EI}\right) .$$

Now we know the change in surface energy during crack extension and the change in elastic strain energy during crack extension. For the crack process to occur, these quantities must balance, or

$$\delta U = \delta S$$

$$2\left(\frac{P^2L^2\delta L}{EI}\right) = 2\delta Lb\gamma,$$

or

$$\frac{P^2L^2}{EI} = b\gamma.$$

If the beam sections are rectangular, then

$$I = \frac{bh^3}{12}$$

and

$$\frac{12P^2L^2}{Eb^2h^3} = \gamma.$$

Here, the aforementioned is solved for the surface energy, γ. For this case, for a given load and geometry, this is the critical surface energy necessary to avoid fracture. Likewise, for a known load and surface energy, the critical crack length, L, can be solved, or, for a known crack length and surface energy, the critical load to initiate fracture, P, can be determined.

Two cantilever beams, however, are an unlikely structural component. Regardless, the same consideration and energy balance can be used for any fracture problem—equating the change in elastic strain energy with the change in surface energy during crack extension. A. A. Griffith recognized this in the 1920s, and the general approach is one of the most critical developments in materials science in the twentieth century.

Griffith Criteria

Griffith considered the fracture process as a simple energy balance. Griffith first considered the general elastic strain energy for a linear-elastic material, which we already know to be

$$U = \frac{1}{2}\sigma\epsilon V = \frac{1}{2}\frac{\sigma^2}{E}V.$$

Then consider a material specimen subject to stress when a crack extends a depth "*a*," the region of the material adjacent to the free surface is unloaded (carries no stress), and the strain energy is released. This is a bit of a complex problem, but can be easily visualized (slightly modifying **figure 5.4**, as shown in **figure 5.7**).

It turns out that the height, "*h*," of those triangles can be approximately set equal to "πa" (or, more accurately, the exact solution for the unloaded area is equal to "$\frac{1}{2}\pi a^2$"; it is not exactly

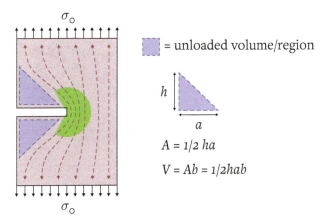

= unloaded volume/region

$A = 1/2\ ha$

$V = Ab = 1/2hab$

FIGURE 5.7   Stress-free material regions (triangles) after crack propagates/extends to a depth "*a*." The material cannot carry stress because of the free surface.

a triangle). The strain energy released (i.e., the difference in strain energy between the uncracked system in **figure 5.4** and cracked system in **figure 5.7**) is then simply the strain energy that was in those unloaded areas, or

$$\delta U = 2 \times \frac{1}{2}\frac{\sigma^2}{E}\left(\frac{1}{2}\pi a^2 b\right) = \frac{1}{2}\frac{\sigma^2}{E}\pi a^2 b,$$

where "*b*" is the material thickness. We see that the change in strain energy is proportional to the crack size, where $\delta U \propto a^2$.

Our next consideration, like the beam example, is the energy necessary to create two surfaces. This term is expressed as

$$\delta S = 2ab\gamma,$$

where "*a*" is the crack length, "*b*" the material thickness, and "γ" the surface energy. Here the change in surface energy is also proportional to the crack size, where $\delta S \propto a$. The total energy associated with a crack is thus the sum of the energy required to create two free surfaces (positive) and the strain energy released in the unloaded portions of the material (negative). These relations are plotted in **figure 5.8**.

As a crack grows longer (*a* increases), the quadratic dependence of strain energy dominates the surface energy; that is, total system energy can be lowered by letting the crack grow longer (see **figure 5.8**). This point defines the critical crack length, a_c,

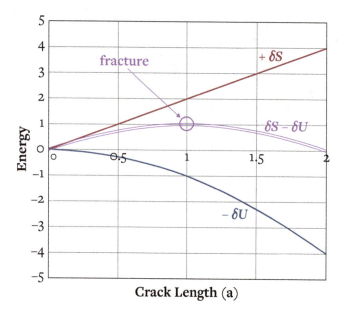

for a given stress and is the maximum of the total energy. We can compute this using a simple derivative:

$$\frac{d}{da}(\delta S - \delta U) = 2b\gamma - \frac{\sigma^2}{E}\pi ab = 0,$$

$$2\gamma b = \frac{\sigma^2}{E}\pi ab,$$

$$\sigma_f = \sqrt{\frac{2E\gamma}{\pi a}}.$$

FIGURE 5.8 Relationship between surface energy (δS) and elastic strain energy (δU) for an idealized system as a function of crack length (a). Fracture occurs at the minimization of free energy ($\delta S - \delta U$).

As it turns out, however, depending on the material, 100 percent of the available energy is not involved in surface creation (there could be energy dissipation due to plastic flow, for example). Irwin and Orowan did this work in the 1940s. The adjusted criteria account for a more general energy requirement than surface energy, encompassing all energy dissipating mechanisms. They suggested that catastrophic fracture occurs when the critical strain energy release rate, G_c, is attained, where

$$\sigma_f = \sqrt{\frac{EG_c}{\pi a}}.$$

The aforementioned is known as Griffith's fracture criteria, relating fracture strength (σ_f) to material stiffness (E) and critical energy release (G_c), and critical crack length (a). Depending on the application, any of the parameters can be of importance. Of note, there are two material-dependent values: E and G_c. Since, for materials such as steel or titanium, these values are constant, they can, in theory, be easily combined into a single constant when considering fracture; this term is known as fracture toughness, commonly denoted as K_{IC}, and will be discussed in the next section.

Fracture Modes and Formula

Alternative to energy balance, fracture can also be considered with respect to the amplified stress intensity at a crack tip. These solutions are highly complex and require known geometric, loading, and boundary conditions. There are three types of possible cracks considered for fracture. We have been considering mode I.

Considering mode I, the stress in the x-direction close to the crack tip can be calculated as

$$\sigma_x(r,\theta) \cong \frac{K_I}{\sqrt{2\pi r}} \cos\frac{\theta}{2}\left(1 - \sin\frac{\theta}{2}\sin\frac{3\theta}{2}\right),$$

which holds for distances close to the crack tip ($r < 0.1a$). The K_I parameter is very important and known as the stress intensity factor (the subscript "I" indicates mode I). As can be seen in the earlier relation,

$$\lim_{r \to 0} \sigma_x = \infty.$$

That is, the stress approaches infinity as the crack tip is approached. Clearly, this is not a physical result. The K_I factor contains the dependence on the applied stress, the crack length, a, and the specimen geometry. For example, for the specific case of a central crack of width $2a$, $K_I = \sigma_0\sqrt{\pi a}$. Some examples of K_I factors are given in **table 5.7**.

These stress intensity factors are easily applied in design by considering the material can withstand crack-tip stresses up to the specified limit, or critical stress intensity, K_{IC}, beyond which fracture occurs. The critical stress intensity factor, K_{IC}, is a measure of material toughness.

The failure stress, σ_f, can then be related to the crack length, a, and the fracture toughness by

$$\sigma_f = \beta \frac{K_{IC}}{\sqrt{\pi a}},$$

TABLE 5.7 EXAMPLE STRESS INTENSITY FACTORS FOR COMMON CRACK GEOMETRIES

Type of Crack	Stress Intensity Factor, K_I
Center crack, length $2a$, infinite plate	$\sigma_0\sqrt{\pi a}$
Edge crack, length a, semi-infinite plate	$1.12\sigma_0\sqrt{\pi a}$
Central penny-shaped crack, radius a, infinite body	$2\sigma_0\sqrt{\dfrac{a}{\pi}}$
Central crack, length $2a$, plate of finite width W	$\sigma_0\sqrt{W\tan\left(\dfrac{\pi a}{W}\right)}$

TABLE 5.8 EXAMPLE FRACTURE TOUGHNESS OF MATERIALS

Material	K_{IC} (MPa-m1/2)	E (GPa)
Steel alloy	150	210
Copper alloy	95	135
Aluminum alloy	28	70
Titanium alloy	66	100
High-density polyethylene (HDPE)	3.5	0.7
Nylon	3.0	3.0
Tungsten carbide	3.7	550
Wood	2.5	9
Concrete	0.75	48
Ice	0.11	9.1

where "β" is a geometrical constant. Clearly, we see the similarity with Griffith's criteria, where

$$\sigma_f = \beta \frac{K_{IC}}{\sqrt{\pi a}} = \sqrt{\frac{EG_c}{\pi a}},$$

resulting in

$$K_{IC}^2 = \alpha EG_c,$$

where "α" is simply another geometric constant ($\alpha = \beta^{-2}$). Typical values for K_{IC} and E are listed in **Table 5.8**.

V. Fatigue Failure

How do cracks grow and eventually lead to fracture? If a material is subjected to many instances of stress, damage accumulates in a process known as **fatigue**. Related to fracture, fatigue can be defined as the weakening of a material caused by repeatedly applied cyclic loads (i.e., repeated cycles of loading and unloading, such as cars traversing a bridge, the loads on an airplane's wings). Repeated loads—even those well below yield point—lead to progressive and localized structural damage. As a result, the nominal maximum stress values that cause such damage may be much less than the strength of the material (UTS limit or the yield stress limit). Fatigue failure is dangerous because a single (static) application of load will not seem sufficient to cause any damage/failure, and thus conventional stress analysis is not sufficient.

Basically, no material is *perfect*. If the loads are above a certain threshold, microscopic cracks will begin to form at the stress concentrators (i.e., defects), such as the surface and grain interfaces. Eventually, a crack will reach a critical size; the crack will propagate suddenly, and the structure will fracture. The shape of the structure will significantly affect the fatigue life; square holes or sharp corners will lead to elevated local stresses where fatigue cracks can initiate. Round holes and smooth transitions or fillets will therefore increase the fatigue strength of the structure. Particularly, in metal alloys, when there are no macroscopic or microscopic discontinuities, the process starts with dislocation movements, which eventually form persistent slip bands that become the nucleus of short cracks. Unlike elastic behavior, fatigue is characterized by a microscopic change in structure and thus the damage is cumulative; materials do not recover when rested.

The **fatigue life, N_f,** of a material can be defined as the number of stress cycles of a specified magnitude that a specimen sustains before failure of a specified nature occurs. If we know the fatigue life, then we can thus replace/repair a structural component before fatigue failure.

Miner's Rule of Cumulative Damages

Let us start with a simple example to help us explore fatigue life. Metal paper clips can be bent so many times in a certain way before they fail—which we will define as the metal snapping into two. If I pick up a paper clip and bend it open 90° (as shown in **figure 5.9 a–c**) and then back to its original position, it takes approximately ten cycles to fail the clip. Therefore, the fatigue life for a "stress" of 90° would be $N_f = 10$. Obtaining the corresponding values of stress to cycles to failure can be very helpful at telling us how the paper clip could behave in other stress scenarios as well. One very simple and widely used model that can help us extrapolate how long the paper clip will stay together is called the Miner's Rule of Cumulative Damages. This model is expressed by the following equation:

$$\sum_{i=1}^{k} \frac{n_i}{N_f} = C,$$

where

- n_i = number of cycles accumulated at a stress σ_i
- N_f = the average number of cycles at a stress level σ_i that causes a failure
- k = the number of stress levels
- C = fraction of life consumed by exposure to the cycles at the different stress levels
- when the damage fraction $\rightarrow 1$ failure occurs

Continuing with our paper clip example, we can look at a series of different stresses and use Miner's Rule to make some estimates of how much of a paper clip's "life" has been used and how much it could be loaded further. Let's start by assuming that the previous stress of 90° done to the paper clip for ten cycles is

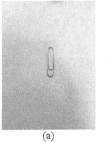  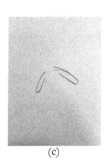

(a) (b) (c)

FIGURE 5.9 (a) Unbent paper clip, (b) paper clip open to 90°, and (c) paper clip at failure.

representative of all paper clips in a given box. We can then say that the work required to fail a given paper clip, $W_{failure}$, is given by the following:

$$W_{failure} = (N_f)(\sigma_i).$$

So, in this case,

$$W_{failure} = (10)(90°) = 900°.$$

Now that we understand the amount of work it takes to fail a paper clip, we can use Miner's Rule to examine other scenarios. What if we only bent a paper clip to 30°? How many cycles would it take before failure? Using Miner's Rule, we can set up the following expression:

$$\sum_{i=1}^{k} \frac{(n_i)(\sigma_i)}{W_{failure}} = C = \frac{(n_i)(30°)}{900°} = 1 .$$

Solving for n_i, we see that we would have to bend the paper clip 30 times at a stress of 30° to fail the paper clip. We can even take this further and introduce a mix of stress scenarios. Let us take the loading proposal in **table 5.9** and predict the fraction of life consumed for a paper clip:

Given this proposal, we can set up the following equation:

$$\sum_{i=1}^{k} \frac{(n_i)(\sigma_i)}{W_{failure}} = C = \frac{7(15°) + 1(45°) + 2(90°) + 2(105°)}{900°} = \frac{540}{900} = 60\% .$$

Thus we could say that 60 percent of the paper clip's potential for work has been used or (for those who see the glass as half full) the paper clip has 40 percent of its potential for doing work remaining!

S-N Curve

As shown in the previous paper clip example, if we want to know the fatigue life for a given stress level, the most direct way to determine the critical number of cycles is through brute force: load the material cyclically until it fails. Before microstructural understanding and modeling was developed, this was the only means to determine the fatigue life.

In high-cycle fatigue situations, a material's performance is commonly characterized by an *S-N* curve (also called a *Wöhler*

TABLE 5.9 PROPOSED LOADING SCHEDULE FOR BENDING A PAPER CLIP

Stress (degrees)	Number of Cycles
15	7
45	1
90	2
105	2

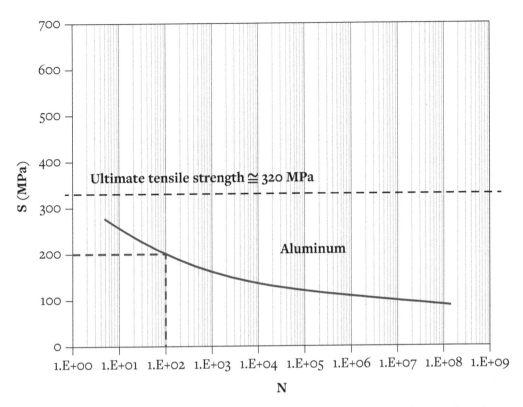

FIGURE 5.10   S-N curve for aluminum (ultimate tensile strength or UTS of approximately 320 MPa). As the cyclic stress amplitude increases, the number of cycles before fracture decreases. For this example, a stress amplitude of 200 MPa would result in a fatigue life of N_f = 100 cycles. For example, the material would fracture after 100 loadings. With increasing cycles (N), we see a continuous decrease in the failure stress amplitude.

curve; see **figure 5.10**). To carry out the test, a cyclic stress of amplitude/magnitude S is applied to a specimen and the number of loading cycles N until the specimen fails is determined. Then S is varied (increased or decreased) and the test is repeated. Since millions of cycles (or more) might be required to cause failure (especially at lower stresses), the stress is plotted against the logarithmic scale of cycles to failure (N).

S-N curves are derived from tests on samples of the material to be characterized where a regular sinusoidal stress is applied by a testing machine that also counts the number of cycles to failure. Each test generates a point on the plot, although in some cases, there is a _runout_ where the time to failure exceeds that available for the test. Clearly, obtaining a full S-N curve is a tedious and expensive process. In addition, fatigue is a process that has a degree of randomness (stochastic), often showing considerable scatter even in well-controlled environments. Moreover, the progression of the _S-N curve_ can be influenced by many factors, such as corrosion, temperature, residual stresses, and the presence of notches.

In some materials, such as ferrous alloys (e.g., steel), the S-N curve flattens out, such that below a certain endurance limit, failure does not occur, regardless of the number of cycles (see **figure 5.11**). The fatigue limit, endurance limit, and/or fatigue strength are all expressions used to describe this property of materials. Ferrous alloys and titanium alloys have a distinct limit, an amplitude below which there appears to be no number of cycles that will cause failure. Other structural metals, such as aluminum and copper, do not have a distinct limit and will eventually fail even from small stress amplitudes. In these cases, a number of cycles (usually 10^7) is chosen to represent the fatigue life of the material.

Assuming Risk: Understanding the S-N-P Curve

One major drawback of the S-N models and the Miner's Rule is that they rely on average values that do not account for the probability of failure that may be unique to a given

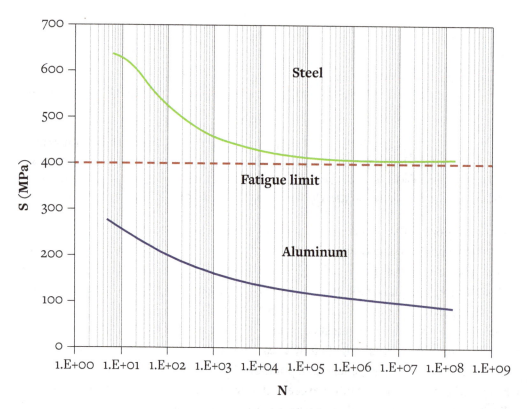

FIGURE 5.11   S-N curves for aluminum and steel. For aluminum, we observe that as the number of cycles (*N*) increases, there is a continuous decrease in the stress amplitude for failure. For steel, when the stress amplitude is less than 400 MPa, no failure occurs. In this case, this is known as the fatigue limit, endurance limit, or fatigue strength of the material.

material. Factors such as corrosion, temperature, and manufacturing defects can all come into play when we do not have the creature comforts that are afforded in lab testing. As civil engineers, we should expect that a material will not be performing in perfect lab conditions. So, what can we do to make our S-N curves more reflective of the risks so that we can take the proper precautions when designing? Include probability in our models, of course!

While it is not in scope of this book to teach probabilistic models in-depth, we do want to emphasize that engineers should be able to take testing data and use it to apply judgment for decision making for materials. The following example is offered to help provide a bit of breadth in how decisions can be made in selecting materials.

During lab testing, it was found that the average minimum fatigue life for a new rivet material used to connect the fuselage in aircrafts was 90 min with a standard deviation of 13.5 percent. According to Federal Aviation Agency (FAA) regulations, if the minimum fatigue life is less than 50 min, the rivet material fails the test. Since the minimum fatigue life of rivets can be said to follow a two-parameter Weibull distribution ($\varepsilon = 0$), what is the probability that this material will fail FAA testing standards?

First, the shape parameter, k_1, of a Weibull distribution is obtained from tables using the standard deviation: $k_1 = 7.91$.

Then the scale parameter, u_1, can be calculated from the average minimum failure, μ_x, and the shape parameter, k_1:

$$u_1 = \frac{\mu_x}{\Gamma\left(1 + \dfrac{1}{k_1}\right)} = \frac{90}{0.93886} = 95.6.$$

Lastly, we solve for the probability that a rivet has a fatigue life less than 50 min and thus the probability of failing the FAA standards test using the Weibull distribution:

$$P(x < 50) = 1 - e^{\left(\frac{-x}{u_1}\right)^{k_1}} = 1 - e^{\left(\frac{-50}{95.6}\right)^{7.91}} = 0.00591 = 0.591\%.$$

Understanding how often this material would fail under pressure gives engineers a better understanding of risk involved in using a material. Given that many materials used in civil engineering design have risks spelled out in the local and state codes, we can make choices in materials given these standards and use these values of risk to assure clients of the safety of our designs.

To help you visualize how a probability model is useful to estimating risk, let us revisit our paper clip example one last time. If you recall, we estimated our work-to-failure value,

$W_f = 900°$ from Miner's Rule—a great back-of-the-envelope tool (see **figure 5.12a**). Remember that the value used to represent the work to fail the paper clip was based on the average from a certain number of tests done at a stress of 90°. It rather falsely leans on the belief that this value is somehow absolute and can be extrapolated in any direction for a given material—sometimes it can, but not always! It should also be noted that while some materials or systems will follow a normal distribution of failure, others may be better modeled with distributions such as Weibull, Poisson, or Gaussian. When making serious predictions of a material's behavior, it is important to include a level of risk or uncertainty with your calculations. If we actually took many paper clips and brought them to the lab, where we could control our stresses, we would then have our S-N curve (see **figure 5.12b**). If you look carefully, the S-N curve has a slight variability as compared to the Miner's Rule curve. Lastly, let us take the 5° stress data from the S-N curve and run a Weibull power distribution on it to create an S-N-P curve (**figure 5.12c**). From Miner's Rule, we can predict at 5° stress that we would have

$$\sum_{i=1}^{k} \frac{(n_i)(\sigma_i)}{W_{failure}} = C = \frac{n_i(5°)}{900°} = 1 \rightarrow n_i = \frac{900°}{5°} = 180.$$

However, using the S-N-P curve for 5°, we can say with confidence that 90 percent of paper clips will make it to 140 cycles (for those who see the glass as half full), or 10 percent of paper clips stressed at 5° will not make it past 140 cycles (as indicated by the star on **figure 5.12c**).

Paris's Law

It is vital that engineers and/or materials scientists are able to predict the rate of crack growth during load cycling such that possible fracture can be curtailed before the crack reaches a critical length, and catastrophic failure can be avoided. A large body of empirical observation suggests that the crack growth can be correlated with the variation in stress intensity factor. This relation is also known as Paris's law.

Paris's law (also known as the Paris-Erdogan law) relates the stress intensity factor range, ΔK, to subcritical crack (nonfracture) growth under a fatigue stress regime; that is, the crack growth with number of cycles, N (see **figure 5.13**). As such, it is the most popular **fatigue crack growth model** used in materials science and fracture mechanics. The basic formula is

$$\frac{da}{dN} = C\Delta K^m,$$

where a is the crack length and N is the number of load cycles, C is a fitted material parameter/constant, and m is a fitted parameter/constant, typically in the range three to five

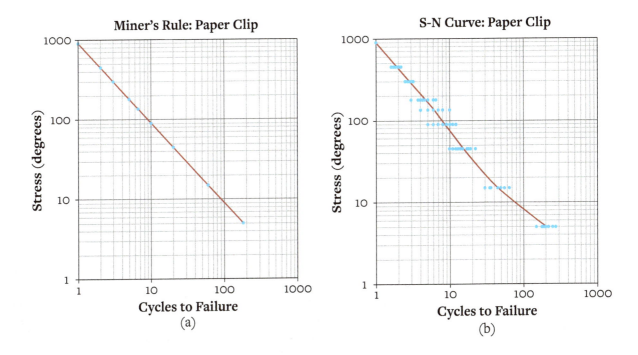

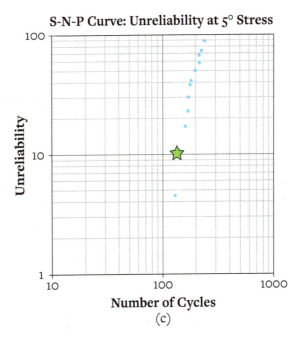

FIGURE 5.12   (a) Miner's Rule for a paper clip, (b) S-N curves for paper clip testing, and (c) S-N-P curve for a paper clip stressed at 5°.

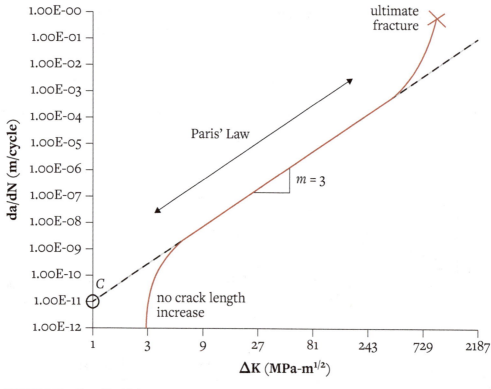

FIGURE 5.13 Plot of Paris's law.

(for metals). Finally, ΔK is the range of the stress intensity factor (i.e., the difference between the stress intensity factor at maximum and minimum loading):

$$\Delta K = K_{max} - K_{min},$$

where K_{max} is the maximum stress intensity factor and K_{min} is the minimum stress intensity factor. The greater the applied stress range, the shorter the life.

We note that Paris's law is purely empirical; that is, there is no fundamental mechanistic reason for the fitted exponent, m.

VI. Concluding Remarks

In this chapter, we discussed the importance of metals as engineering materials and in particular the production and treatment of steel. In terms of mechanical behavior, the occurrence of fracture and fatigue are unique to metal systems and should be considered for any civil engineering design using such materials.

Note that this material is introductory only. In materials science, entire texts are dedicated to the fields of metallurgy, alloys, and steel. Fracture mechanics is a subject unto itself. We did not even discuss plasticity, temperature effects, fire resistance, or technological advancements in processing metals (e.g., control of grain boundaries, porosity, metal matrix composites). However, an appreciation for the unique behavior of metals and steel is necessary for a well-rounded civil engineer, regardless of deep knowledge of particular idiosyncrasies.

VII. Problems

1. Describe the difference between annealing and tempering for steel. Address the outcome of each process on the metal properties in your answer.

2. Explain how carbon content influences the mechanical properties of steel.

3. What causes a fracture to go critical? Explain the process.

4. Explain how fatigue can lead a bridge to fail at a load well below its yield point.

5. How much would a 5-foot section of a HP14 × 102 beam weigh?

6. Steel has a modulus of elasticity of about 30×10^6 psi and aluminum about 10×10^6 psi. When a tensile load of 400 pounds is applied to specimens of equal cross-sectional areas, one steel and one aluminum, which will deform more? Explain with evidence.

7. Given an axial load of 200 N, determine the stress and deformation in an aluminum wire 20-meters long and 5-millimeters wide.

8. An aluminum alloy specimen with a radius of 0.28 inches was subjected to tension until fracture and produced the following results:

Stress, ksi	Strain, 10^{-3} inches/inch	Stress, ksi	Strain, 10^{-3} inches/inch
8	0.6	62	5.8
17	1.5	64	6.2
27	2.4	65	6.5
35	3.2	67	7.3
43	4.0	68	8.1
50	4.6	70	9.7
58	5.2		

a. Using a spreadsheet program, plot the stress-strain relationship.

b. Calculate the modulus of elasticity of the aluminum alloy.

c. Determine the proportional limit.

d. What is the maximum load if the stress in the bar is not to exceed the proportional limit?

e. Determine the 0.2 percent offset yield strength.

f. Determine the tensile strength.

g. Determine the percent elongation at failure.

9. A large steel gusset plate (K_{IC} = 50 MPa-m$^{0.5}$) in a truss bridge undergoes cyclic *tensile* (300 MPa) stresses during use. Prior to use, it was inspected using ultrasonic techniques, from which the largest surface crack found was 2.5 millimeters in length (minimum crack length once loading begins). For the steel in question, the Paris law constants are

$$C = 1.5 \times 10^{-12} \text{ m/(MPa-m}^{0.5}) \text{ per cycle}$$

$$m = 2.5$$

Calculate the number of cycles to failure, N_f. Take the stress intensity factor, $K = \sigma\sqrt{\pi a}$. Note: Integrating Paris's law yields

$$N = \int_{min}^{max} \frac{1}{C\Delta K^m} da.$$

Hint: First determine the maximum crack length from the fracture strength, then use the change in actual stresses to determine ΔK as a function of crack length and solve for the critical number of cycles based on the minimum and maximum crack length.

10. A steel specimen is cyclically loaded in tension to a stress of 15 ksi and fails after 2,000 cycles. A structural member of the same steel is loaded in tension to 20 ksi. Using Miner's rule, how many cycles of this loading can it withstand before failure?

11. A round bar of A36 steel has a length of 2 feet and diameter of 0.25 inches. The stress strain curve for this material is given by the equation

$$\varepsilon = \frac{\sigma}{30,000}\left[1+\frac{3}{7}\left(\frac{\sigma}{50}\right)^5\right]$$

in which the units for σ are kips per in² (ksi). You decide to strain harden the material by loading to a strain $\varepsilon = 0.015$. Assume that the slope of the unload-reload line is the same as the <u>initial</u> tangent modulus.

    a. What is the initial tangent modulus for the material? Use the exact method (i.e., take the derivative).

    b. After the strain hardening and unloading sequence, what is the new yield strength of the material, using the 0.1 percent offset? You should graph this using a spreadsheet, with stress increments of 5 ksi up to 95 ksi.

12. The figure shows an aluminum bar with cross-section 3 in² is surrounded by a steel sleeve that has a cross-section of 1 in² and the same length as the aluminum bar. What is the maximum compressive load that can be supported by this composite piece without either metal yielding? For aluminum, E = 10 × 10⁶ psi and yield = 47,000 psi; for steel, E = 30 × 10⁶ psi and yield = 70,000 psi. Note that the strain will be the same in both metals.

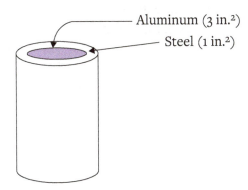

Aluminum (3 in.²)
Steel (1 in.²)

13. Consider the flat, cracked, rectangular plate, as depicted, which has a thickness "t." The plate will be loaded by a tensile force, P, which acts perpendicular to the crack. For this situation, the strain energy is given by the following equation (where P = applied force and L = crack length):

$$U(L) = \frac{1}{2}\sigma\varepsilon V = \frac{\sigma^2 V}{2E} = \frac{\left(\frac{P}{A}\right)^2 (A \cdot L)}{2E} = \frac{P^2 L}{2EA}.$$

If force, P, causes the crack length to increase by ΔL, the strain energy with the newly cracked material is written as follows:

$$U(L+\Delta L) = \frac{P^2(L+\Delta L)}{2EA}$$

a. Determine a simplified expression for the difference in strain energies (elongated crack versus initial crack).

b. Determine a simplified expression for the change in surface energy, assuming the plate material has an adhesion energy per length, γ, in J/m and a thickness of 1.

c. Solve to determine an expression for the required force, P, to elongate the crack.

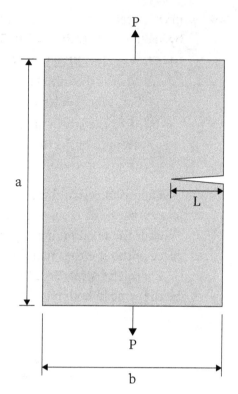

6

Aggregates and Cementitious Materials

I. Introduction

For over 2.5 million years, humans have used stone for shelter and various tools. Early humans inhabited caves and even carved homes into the soft volcanic rock with their hands (see **figure 6.1a**). With the advent of metallic tools, civilizations were able create stone blocks and even get creative with their architectural styling (see **figures 6.1b and 6.1c**). Early civilizations built grand monuments to showcase their talent with stone. The grandiose nature of building large stone structures also made way for advances in math and architectural physics, advancing civilizations. This sustainable material has held up better than any construction material thus far and is still valued in construction today for its strength and architectural splendor.

The one downside to stone construction is that it's pretty heavy to move large blocks and time consuming to shape, so humans got creative. Approximately 8,500 years ago, humans discovered the benefits of hydraulic lime mixed with water to form a cementitious material. These early structures enabled construction of small housing units and underground water cisterns, allowing many early trading civilizations to flourish. Later, the Romans experimented with cementitious materials to create elegant shapes, such as arches, in their structures. They were known for mixing volcanic ash with sea water for strength, horsehair for reinforcement, and even blood for frost resistance. The experimentation with cementitious materials continues today as we push the known boundaries of structural and architectural possibilities. The buildings of the world are getting taller due to new reinforcement techniques and a firmer grasp on the chemistry involved in creating stronger cementitious bonds in concrete. With technology, what we can say about the oldest and arguably most sustainable building material in the world is what's old is new again.

II. Aggregate Sources

An **aggregate** is a combination of inert, granular, and inorganic materials. Common aggregates used by civil engineers include soil, stone, or stone-like solids. With the ability to select natural or synthetic aggregates, civil engineers have many uses for aggregates. **Table 6.1** provides a few examples of aggregate use in the civil engineering subdisciplines.

Natural aggregates are extracted from stone and soil deposits. Before a natural aggregate arrives to a job site, the original rocks must be crushed, sized, and sorted for specified need and usually washed to remove dust and organics. Natural rock aggregates belong to one of three types, depending on the process in which they were created, which include igneous, sedimentary, and metamorphic rock.

(a)

(b)

(c)

FIGURE 6.1 (a) Cave homes in Cappadocia, Turkey; (b) architectural column at the Basilica Cistern; (c) horseshoe arches at the Hassan II Mosque.

Igneous rocks (fun fact: *Ignis* is Latin for fire) are produced from volcanic action when molten material (magma) cools and hardens. Within the category of igneous rocks, there are two types depending on where the molten material was cooled: intrusive and extrusive. **Intrusive** igneous rocks are formed under the earth's crust from cooling molten magma. The cooling time for intrusive igneous rocks is very long and leads to very large "coarse grains" that can be seen easily by the eye. Intrusive rocks are most likely to find their way to the earth's surface as mountain ranges and through erosion. Examples of intrusive igneous rocks include granite, diorite, and gabbro. Known for its durability, granite is one of the most popular igneous rocks that is used for both structural and architectural components in buildings. It has low reactivity to acid rain (very inert), low water absorption, and high hardness. **Extrusive** igneous rocks are formed at the surface of the earth's crust, where the magma becomes known as lava. The molten material is able to cool quickly, sometimes with air voids and suspended crystals making their way into the structure. This quick cooling and crystallization process leads to "fine-grained" rock. Some examples of extrusive igneous rock include basalt, pumice, and obsidian. Recognized

TABLE 6.1 AGGREGATE IN CIVIL ENGINEERING

Subdiscipline	Examples
Construction	Concrete
	Plaster
	Grout
Transportation	Asphalt cement
	Railroad ballast
	Road subgrades
Environmental	Filters
	Drainage
Geotechnical	Foundation bases
	Retaining wall backfill
	Mechanically stabilized earth

for its hardness, basalt is a common rock used in the construction of cobblestone streets, railroad ballast, and roadbed bases.

Sedimentary rocks are formed when stratified soil and/or organic layers are deposited (usually by flowing water) and then later cemented together by compaction and the pressure of being buried. This type of rock is prevalent only on the surface of the earth, and it is the "softest" rock type of the three, with Mohs Harness values typically between two and three (**table 6.2**). Some of the best-known examples of sedimentary rock include coal (yes, the stuff we burn for energy), gypsum, and limestone. Sedimentary rock is one of the most important to the civil engineers for acquiring sand and gravel used in concrete and asphalt.

Metamorphic rocks form when igneous or sedimentary rocks are exposed to extreme heat and/or pressure, which cause changes in texture and chemical packing known as recrystallization. Well-known examples of this

TABLE 6.2 MOH'S RELATIVE HARDNESS SCALE

Mineral	Moh's Relative Hardness	Reference Object
Talc	1	Plastic bag (1)
Gypsum	2	Fingernail (2.5)
Calcite	3	Penny (3.5)
Fluorite	4	Wire nail (4.5)
Apatite	5	Glass (5.5)
Orthoclase	6	Steel file (6.5)
Quartz	7	Emerald (7.5)
Topaz	8	Masonry drill bit (8.5)
Corundum	9	Ruby (9)
Diamond	10	

FIGURE 6.2   Crushed toilets—recycled for aggregate use!

type of rock include marble, gneiss, slate, and gemstones (like diamonds!) These rocks are extremely hard (six and above on the Moh's scale) and can be used by civil engineers as cutting tools or paving blocks.

Synthetic or artificial aggregates are an upcoming source for aggregates and allow engineers to reuse materials from other processes or use engineered aggregates. **Byproducts of industrial waste** are one type of aggregate that one might find in concrete. A stony waste material from steel production called slag is commonly used in concrete to minimize the amount of Portland cement needed in a mix. Iron ore or steel balls, bars, and punchings left over from steel manufacturing are also great aggregates to include in concrete that can also provide radiative protection. **Recycled waste materials** are another type of synthetic or artificial aggregate. Previously used concrete, asphalt, tire chips, glass, and even crushed toilets (**figure 6.2**) can make great aggregates, bringing new properties to materials such as aesthetics and flexibility. The last type of synthetic or artificial aggregate is **engineered aggregates**. This type of aggregate is finding utility for creating lighter-weight concretes and asphalts. Some examples include heat- and water-expanded clays, Styrofoam beads, and micro-sized glass air bubbles encased in resin.

III. Aggregate Properties

An aggregate's **texture**, **shape**, **durability**, and **toughness** will play a large role in how suitable it will be for various applications. An aggregate's texture, durability, and toughness depend largely on *mineral composition*. As described earlier in this chapter, rock aggregate's mineral composition is based on how it was formed and where certain formations lead to different strength properties (see Moh's hardness, for example). While intrinsic mineral properties can dominate some performance properties, *processing* techniques can also control suitability by altering an aggregate's shape, texture, and durability (depending on the process; see **figure 6.3**). Aggregate processing varies and can range from a natural state to an aggregate that was blasted and then received multiple impact-crushing cycles to achieve the desired properties.

FIGURE 6.3   (a) Angular aggregate particles; (b) round aggregate particles.

The main considerations for **shape** when selecting an aggregate are the **angularity** and **flakiness**. An aggregate's degree of angularity is defined by the number of faces that are fractured, where fractured faces are defined by an angular, rough, or broken surface of an aggregate particle created by crushing, by other artificial means, or by nature. The degree of flakiness is measured by the ratio of an aggregate's thickness to size and helps predict an aggregate's predisposition for future fracture. Quantifying both angularity and flakiness are necessary because they are important for understanding the compactability of an aggregate (see **figure 6.4**).

Linked to shape, but markedly different, is **texture**. The main thing to note when selecting an aggregate for texture is whether it has **angular** or **rounded** edges. As described earlier, angular edges have a *rough* texture because of the nature of their formation (fracturing). Rounded edges can be *smooth* and even appear *polished*, depending on the method in which they were processed. Quantifying an aggregate's texture is important, because it is an indicator of its shear strength. For example, angular aggregate is desirable for ensuring that there is *interlocking* between aggregate pieces under loading conditions. Conversely, aggregates

FIGURE 6.4   (a) Natural gravel; (b) crushed stone; (c) flaky stone.

FIGURE 6.5   Los Angeles abrasion test.

with a rounded texture are more likely to see *rolling* and *sliding* among particles.

The **durability** (soundness) of an aggregate is defined as its resistance to deteriorating agents, ranging from **physical** to **chemical**. A physically durable aggregate is able to resist damages, such as fractures. A chemically durable aggregate must be resistant to reactions with common environmental elements (i.e., be **inert**). The main sources of instability over time that durability aims to avoid include volume changes owing to water absorption, volume changes owing to chemical reaction (oxidation, hydration, carbonation), freeze-thaw cycle damage, and general chemical degradation.

Toughness, hardness, and abrasion resistance are all terms that describe an aggregate *surface's* ability to resist scratching, disintegration, and general degradation. Because there can be mechanical breakdown during shipping and handling, tests such as the **Los Angeles abrasion test** are performed to quantify the degree toughness to which one can expect an aggregate to perform. These tests roll the sediment with standardized metal balls (as shown in **figure 6.5**) and measure the amount of fine sediment produced.

This empirical test compares your aggregate's results to other known results and determines a toughness. By choosing an aggregate with an improper toughness, unwanted dust and small rock fragments can lead to both physical and chemical detriments. While physical dust is an obvious environmental hazard (causing breathing issues and sediment pollution in waterways), excess fine particles can also alter the results of chemical processes, such as concrete curing times and ultimate strength.

Quantifying Aggregate Moistures Properties

When quantifying the physical and chemical properties of an aggregate, what is present is just as important as what is missing. The absence of matter in aggregate means that space has the capability to be filled. In fact, empty space even has its own absorptive physics dependent on how small the space is between matter (think about how the meniscus travels up a tube!). With the knowledge that empty spaces have forces of their own, it is important to identify and differentiate these spaces within an aggregate. **Pores** are the spaces *inside of an aggregate* that can include anything from microcracks to macro-sized openings (e.g., pumice rock). Pores belong to individual aggregate pieces and usually cannot be increased or decreased. **Voids** are the spaces between pieces of aggregate, and their sizes can easily be changed by altering the packing of an aggregate. Lastly, the **solids** portion of the aggregate is inclusive of the continuous matter that makes up the aggregate. With an understanding of basic aggregate nomenclature (see **figure 6.6**), we are able to further describe an aggregate's abilities with respect to how it will behave as an engineered material.

Voids content (v) is the ratio of volume of space between particles to the total volume of an aggregate sample, where

$$v = \frac{V_{voids}}{V_{Total}} \times 100\%.$$

Typical values range from 30–50 percent.

The **porosity** (p) of an aggregate is the measure of volume in each aggregate piece not occupied by solids and is given by

$$p = \frac{V_{Pores}}{V_{bulk}}.$$

With a numerical understanding of the ratios of solid to empty space, we can then describe how substances, specifically water in most cases, can fill these areas. Aggregate has four generally accepted moisture states. These states from driest to most wet

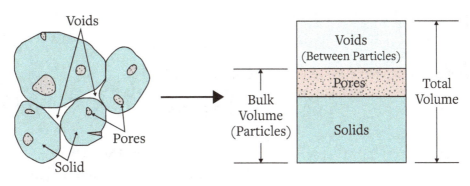

FIGURE 6.6   Basic aggregate nomenclature.

include oven dry (OD), air dry (AD), saturated surface dry (SSD), and wet or moist. While an understanding of all states is necessary, *only OD and SSD can be used reliably* because of their repeatability in lab testing. The definitions of each state are provided in **table 6.3**:

TABLE 6.3 AGGREGATE MOISTURE CONDITIONS

Moisture Condition	Abbreviation	Description	Condition of Pores
Oven dry	OD	All moisture removed by heating	Empty
Air dry	AD	No surface water	Partially full
Saturated surface dry	SSD	No surface water	Completely full
Wet or moist	-	Water film on surface	Completely full

Quantifying these values allows engineers to understand how an aggregate in a particular state will react when introduced to water. This is especially important for mixing concrete when water is needed to facilitate the hydration reaction. If water is in aggregate pores, it is not available for hydration; the desired strength at cure will not be achieved. Therefore, it would make sense to add additional water to account for water absorption into aggregate pores when mixing concrete, dependent on the aggregate's initial moisture state at mix time.

In general, the **moisture content** (**MC**) of aggregate is defined by the ratio of the weight of water contained in the aggregate to the oven dry (OD) weight of the aggregate and is typically expressed as a percent:

$$MC = \frac{\text{Weight of water in aggregate}}{\text{Oven dry weight of aggregate}} (100\%).$$

For OD conditions, all moisture is removed from the aggregate, and therefore the moisture content is zero. If the OD weight of aggregate is known, the weight of water (W_w) in the aggregate may be determined by subtracting the OD weight (W_{OD}) from the current (moist) weight $(W_{current})$ of the aggregate, or

$$W_w = W_{current} - W_{OD}.$$

Therefore, we can write the general MC equation as follows:

$$MC = \frac{W_{current} - W_{OD}}{W_{OD}} (100\%).$$

An aggregate's capacity to absorb moisture, **absorption capacity** (**AC**), represents the quantity of water (moisture) in the aggregate when all the pores are full (i.e., in the SSD condition). AC is determined based on the general MC equation by using the weight of the aggregate in the SSD condition (W_{SSD}) as the current weight. Therefore, AC is given as a percentage by the following equation:

$$AC = \frac{W_{SSD} - W_{OD}}{W_{OD}}(100\%).$$

For determining absorption capacity, the SSD condition is usually achieved by laboratory processes.

In the field, it is difficult to achieve perfect conditions, and there may be excess water on an aggregate's surface or between pores owing to precipitation (e.g., rain, snow). In order to account for this in a concrete mixing process, we must quantify the excess surface moisture (i.e., the amount of moisture not contained in the aggregate pores and therefore available to react with cement in the concrete). This could be achieved by bringing a sample of the wet aggregate to a lab, weighing it, weighing it again after bringing it back to the SSD condition, and then weighing it a third time after drying it to the oven dry condition. The percentage of **surface moisture content (SMC)** would then be given by

$$SMC = \frac{W_{WET} - W_{SSD}}{W_{OD}}(100\%).^{[1]}$$

Put another way, the surface moisture content is equal to the current moisture content of the aggregate, minus its absorption capacity, or

$$SMC = MC - AC.$$

Quantifying Aggregate Weight and Volume Properties

Weight is increasingly becoming a factor in concrete design. Since the majority of the weight in a concrete mix comes from the aggregate, it is important to understand the proportions of an aggregate with respect to weight and volume. Traditional ways of thinking about this relationship are through density (ρ) or unit weight (γ) (discussed in **chapter 1**). However, **specific gravity** (G) is the unitless term that is most commonly used in industry. Specific gravity is defined as

$$G = \frac{\rho_{material\ @\ Temperature\ T}}{\rho_{water\ @\ Temperature\ T}} = \frac{\gamma_{material\ @\ Temperature\ T}}{\gamma_{water\ @\ Temperature\ T}}.$$

---

1 Important point to note for SI units use! Since the equations for moisture content are unitless, mass can be used interchangeably with weight to yield identical values.

EXAMPLE 6.1: MEASURING MOISTURE AND ADJUSTING A CONCRETE MIX

An aggregate weighs 11.43 kilograms in its natural condition, 10.84 kilograms after oven drying, and 11.16 kilograms in its SSD condition.

 a. What is the total MC of this aggregate?

 b. What is the absorption capacity of this aggregate?

 c. What is the SMC?

 d. If a concrete mixture is designed to have 136 kilograms of water added to 1,390 kilograms of dry aggregate, how much water has to be added to SSD aggregate?

SOLUTION

 a.

$$MC = \frac{M_{current} - M_{OD}}{M_{OD}}(100\%) = \frac{11.43\text{kg} - 10.84 \text{ kg}}{10.84 \text{ kg}}(100\%) = 5.44\%$$

 b.

$$AC = \frac{W_{SSD} - W_{OD}}{W_{OD}}(100\%) = \frac{11.16 \text{ kg} - 10.84 \text{ kg}}{10.84 \text{ kg}}(100\%) = 2.95\%$$

 c. Since we know the aggregate's natural condition mass is greater than the SSD, we can conclude that it is wet.

$$SMC = \frac{W_{WET} - W_{SSD}}{W_{SSD}}(100\%) = \frac{11.43\text{kg} - 11.16 \text{ kg}}{10.84 \text{ kg}}(100\%) = 2.49\%$$

Or, you could conclude that the surface moisture is the difference between the total moisture and the absorbed moisture:

$$SMC = 5.44\% - 2.95\% = 2.49\%$$

 d. If we used 1,390 kilograms of dry aggregate, we know from part (b) that it would have absorbed:

$$1{,}390 \text{ kg}(2.95\%) = 41\text{kg of water.}$$

Therefore, if the aggregate is in the SSD condition already, then we only need

$$136 \text{ kg} - 41\text{kg} = 95 \text{ kg of water}$$

for the concrete mix.

When describing aggregate, we need to specify the moisture condition. However, we cannot calculate specific gravity in terms of a moisture condition greater than SSD, because specific gravity *does not take into account void volume*, only solids and pore volume. Thus, specific gravity can be rewritten as the following equations:

$$G_{bulk-dry} = \frac{\text{Oven Dry Weight}}{\text{Bulk Volume}(\gamma_{water})} = \frac{W_{OD}}{V_{bulk} \cdot \gamma_w} = \frac{\gamma_{bulk}}{\gamma_{water}},$$

and

$$G_{bulk-SSD} = \frac{\text{SSD Weight}}{\text{Bulk Volume}(\gamma_{water})} = \frac{W_{SSD}}{V_{bulk} \cdot \gamma_w} = \frac{\gamma_{bulk-SSD}}{\gamma_{water}},$$

and

$$G_{bulk-total} = \frac{\text{Weight between OD and SSD}}{\text{Bulk Volume}(\gamma_{water})} = \frac{W_{Total}}{V_{bulk} \cdot \gamma_w} = \frac{\gamma_{bulk-total}}{\gamma_{water}}.$$

IV. Aggregate Sorting and Classification

Gradation describes the particle size distribution of an aggregate. Knowledge of the distribution of particle sizes within an aggregate sample is important for many design considerations. For example, small particles have low strength and require more binder to hold them together but are easy to move and work into place. Large particles are stronger and require less binder but are harder to transport and work into place. In some cases, large particles would be ideal for strength, but their size is prohibitive due to constraints, such as rebar grid spacing or the dimensions of a particular construction space.

Gradation is measured through a process of sorting aggregate particles. Sieves of various mesh-size openings are stacked from large opening sizes at the top to smaller openings followed by a collection pan at the bottom. An aggregate sample is poured into the top sieve, and gravity and a mechanical shaking machine are used to vibrate increasingly smaller particles downward until they are unable to pass through a sieve opening to their effective diameter. The effective diameter assumes that each aggregate can have an assumed finite spherical size that can be measured through the sieving process by an aggregate's inability to pass a specific sieve opening. At the end of the sieving process, the aggregate sample will be sorted by the effective diameters on each sieve, and a distribution can be plotted.

A grain-size distribution is the most common method of presentation of an aggregate gradation. **Grain-size distributions** depict the percentage by weight (or mass) of aggregate retained on or passing each sieve versus the log of various sieve sizes (or grain-size

EXAMPLE 6.2: AGGREGATES AND DELIVERY TRUCK SIZING

You must deliver 5 tons (1 ton = 2,000 pounds) to a job site. The quarry informs you that the dry bulk specific gravity of this aggregate is 2.79, and its void content is 38 percent when loaded into the truck. What volume in cubic yards (3 feet = 1 yard) is required for the delivery truck?

SOLUTION

From

$$G_{bulk} = \frac{\gamma_{bulk}}{\gamma_{water}},$$

we can rearrange to find the volume of aggregate:

$$\gamma_{bulk} = G_{bulk} \cdot \gamma_{water} = 2.79 \cdot 62.4\frac{lb}{ft^3} = 174.096\frac{lb}{ft^3}.$$

Knowing that we need 5 tons of aggregate delivered to the job site, we can find the volume that just the bulk aggregate will need.

$$V_{bulk} = \frac{\text{Weight aggregate needed}}{\gamma_{bulk}} = \frac{5\text{tons} \times \dfrac{2,000\ lb}{1\ ton}}{174.096\ \dfrac{lb}{ft^3}} = 57.4\ ft^3$$

Now we must calculate the total volume needed for this aggregate, keeping in mind that when it is loaded, there will be a 38 percent void volume. Since the remaining 62 percent of the volume of the truck is bulk solids, we can say that the total truck size needed is

$$V_T = \frac{V_{bulk}}{62\%} = \frac{57.4ft^3}{0.62} = 92.6ft^3 = 3.43\ yd^3.$$

diameters) used to perform the gradation test. A **log scale** on the grain-size diameter is necessary to compare the diameter values collected as they fall over a large range. Without the log scale, the plot would be skewed by large values, and the details of the smaller grains would be less pronounced.

There are many ways to read a grain-size distribution (see **figure 6.7**). The first bit of information to gather from the test result is whether your aggregate has coarse or fine grain sizes. The dividing line between coarse and fine grains is the commonly named #4 sieve, which has an effective diameter of 4.75 millimeters (3.16 inches). **Fine grains**, such as sand,

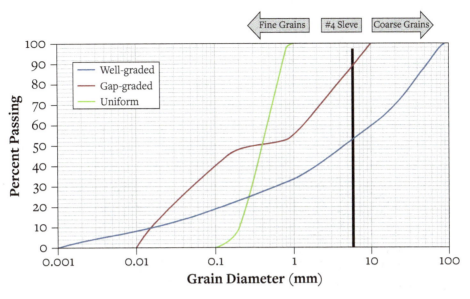

FIGURE 6.7   Grain-size distribution examples.

silt, and clay, are found in the sieve sizes below the #4, and **coarse grains**, such as gravel, cobbles, and boulders, have effective diameters greater than the #4 sieve.

The second bit of information gained by the grain-size distribution is the frequency distribution. The frequency distribution is told through the shape of the plotted grain size distribution curve. A near vertical curve indicates a concentration of particles of that diameter; a near horizontal curve indicates scarcity of particles of those diameters.

Three commonly used terms to describe an aggregate's frequency distribution are uniform, gap graded, and well graded. **Uniform-graded** (also called "poorly graded" or "open-graded") aggregates contain many particles of the same size. A grain-size distribution that is uniform usually depicts a nearly vertical line. **Gap-graded** aggregates have most particles concentrated at two or three diameter sizes. These distributions are noted by a "staircase"-shaped, grain-size distribution. **Well-graded** (also called "continuous" or "dense") aggregates have an even distribution of sizes and are indicated by a continuous sloping line.

Determining an exact value on the x-axis of a log-scale graph can sometimes be very difficult. The value is halfway between two major points such as 0.1, and 1 is not 0.5; linear interpolation rules do not apply to a log scale. Remembering that a log scale allows smaller values to be emphasized, a value halfway between 0.1 and 1 is actually closer to 0.365, and one must use **log interpolation** when finding this value. The best way to find values accurately for log interpolation is to use a ruler to measure distances (see **figure 6.8** in the example that follows).

EXAMPLE 6.3: LOG INTERPOLATION

Let's say we wanted to find the grain diameter for the given distribution shown that is 50 percent finer by weight. Obviously, we could read it from the graph, but an exact value is hard to find. We can obtain a more exact value be implementing the following procedure:

1. *Measure the distance between the major gridlines provided (6.4 centimeters in figure 6.8).*

2. *Measure the distance from one grid line to the point of interest (3.6 centimeters in figure 6.8).*

3. *Set up your interpolation equation:*

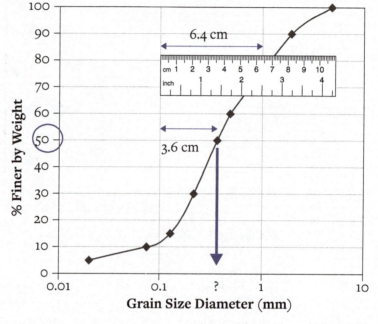

$$\frac{\log x - \log 0.1}{\log 1 - \log 0.1} = \frac{3.6}{6.4}.$$

4. *Solve for x.*

$$x = 10^{\left(\frac{3.6}{6.4} - 1\right)} = 0.365 \, \text{mm}$$

Figure 6.8   Log interpolation example.

TABLE 6.4 FINENESS MODULUS

Range	Classification
0–1.5	Very fine
1.5–3.0	Moderately fine
3.0–4.5	Moderately coarse
4.5–6.0	Coarse

Another measurement that can be taken from the grain-size distribution is the fineness modulus (FM) (see **table 6.4**). The **FM** is an empirical factor that measures the fine aggregate's gradation. Knowing the number of fines in an aggregate is important to know for concrete design and will be discussed later in this chapter. To evaluate

EXAMPLE 6.4: FINENESS MODULUS

Classify the following aggregate based on the results of the sieve analysis shown.

Sieve Number Retained	Sieve Opening Size (mm)	Individual Percent Retained	Cumulative Percent
4	4.75	1	1
8	2.36	18	19
16	1.18	20	39
30	0.6	19	58
50	0.3	18	76
100	0.15	16	92
Pan	–	8	–
		$\sum = 100$	$\sum = 285$

SOLUTION

$$FM = \frac{\sum \text{cumulative \% retained on Sieve Nos. 4, 8, 16, 30, 50, 100}}{100}$$

$$FM = \frac{285}{100} = 2.85$$

Based on FM = 2.85, the aggregate classifies as **moderately fine**.

an aggregate's FM, *six specific sieves* are necessary: the #4, #8, #16, #30, #50, and #100. Using the percent-retained values from these sieves, add cumulative percent retained and divide this value by 100. Once you have this number, you can classify the soil by its FM.

In certain scenarios, you may find that in order to meet the specifications laid out for a particular job, you will have to mix two aggregates together. To figure out the proportions necessary for each aggregate, use the following procedure:

1. Set up a plot with percent passing on the vertical axis, the percent of the total blend for one aggregate along the top horizontal axis, and the percent of the total blend for the other aggregate along the bottom horizontal axis. Example: For aggregates

A and B, the bottom axis can be defined from 0 to 100 percent aggregate A in the blend going from left to right; therefore, the top axis is defined from 100 to 0 percent B going from left to right).

2. Plot the percent passing for aggregates A and B along a vertical line for 100 percent of the particular aggregate in the blend. Example: For aggregate A, 100 percent aggregate A is at the right-hand side of the plot; therefore, plot the percent passing for aggregate A as points along a vertical line at 100 percent aggregate A.

3. Connect the two percent passing points for aggregates A and B for each sieve size; this is your sieve line.

4. Plot the limits allowed in the specifications for each sieve on the sieve lines (i.e., mark two points on each sieve line corresponding to the percent passing limits for that sieve size in the specification).

5. Connect the upper bound points and lower bound points between the sieve lines.

6. Identify the tightest range of vertical lines that can be drawn within the specification limits when accounting for all sieve sizes; these are the allowable bounds for the aggregate blend.

7. Draw a line down the middle of the allowable bounds range; this is the blend line, representing the proportions of aggregates A and B that should be used in the blend.

8. Select the percent passing values from the intersection of the blend line with sieve lines; these represent the grain size distribution of the blended aggregate, based on the proportions of aggregates A and B defining the blend line.

EXAMPLE 6.5: BLENDING AGGREGATES TO MEET SPECIFICATIONS

The table that follows shows the grain-size distribution for two aggregates and the specification limits for an asphalt concrete. Determine the blend proportion of aggregates A and B required to meet the specification and the gradation of the blend.

	Size (mm)								
	19	12.5	9.5	4.75	2.36	0.60	0.30	0.15	0.075
Specification limits (percent passing)	100	90–100	70–90	50–70	35–50	18–29	13–23	8–16	4–10
Aggregate A (percent passing)	100	85	55	20	2	0	0	0	0
Aggregate B (percent passing)	100	100	100	85	67	45	32	19	11

SOLUTION

Steps 1 through 3:

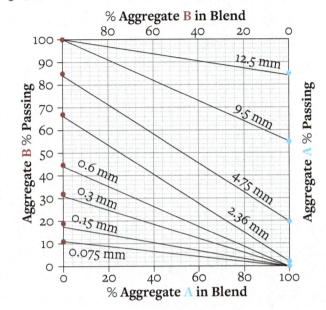

Figure 6.9a   Blending aggregate example, steps 1–3.

Steps 4 and 5:

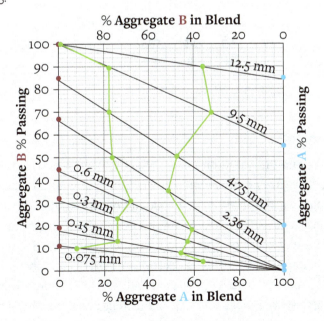

Figure 6.9b   Blending aggregate example, steps 4 and 5.

Steps 6 through 8:

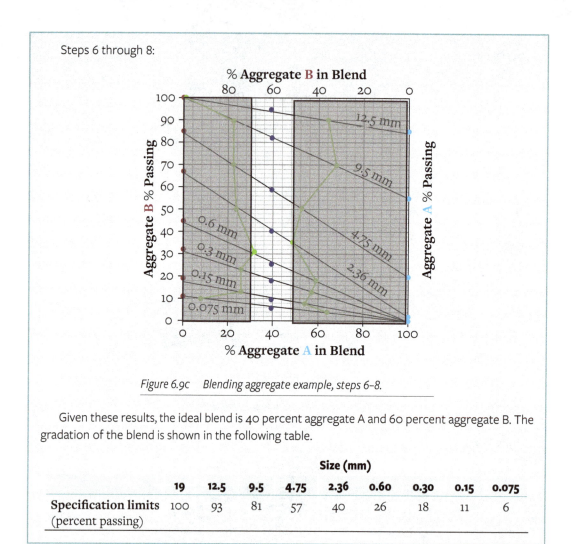

Figure 6.9c   Blending aggregate example, steps 6–8.

Given these results, the ideal blend is 40 percent aggregate A and 60 percent aggregate B. The gradation of the blend is shown in the following table.

	Size (mm)							
	19	12.5	9.5	4.75	2.36	0.60	0.30	0.15	0.075
Specification limits (percent passing)	100	93	81	57	40	26	18	11	6

V. Aggregate Issues

When mixing aggregate into concrete or asphalt, there are a several factors to be aware of to ensure that there are no unwanted results. First, unwanted substances—namely, **absorbent particles**, such as clay, coal, wood, and other organics—can affect the water available to cement particles. This may result in a *decrease in strength* from the desired mix specifications because of a lack of hydration for cement particles. Second, the aggregate's mineral makeup is important to note, as certain minerals can have **adverse reactions** with the alkali-based binders. Particularly, Portland cement is known for its reactivity with silica-based aggregates; *expansion*, *cracking*, and *spalling* are all results of this reaction. It is recommended to either avoid aggregates with these problems or to use an admixture with your concrete to deter the reaction. Lastly, some aggregates have more of an **affinity for water** than others. When mixing asphalt, overly moist aggregate will undergo **stripping**, where asphalt film separates from an aggregate because of moisture.

When handling, storing, and transporting aggregate, the goal is to avoid or minimize *segregation*, *degradation*, and *contamination* for the reasons previously mentioned. To aid in this goal, proper storage of aggregates should include the building of piles in thin layers to avoid larger particles from rolling to the bottom of piles. Piles should also be separated by type (e.g., rounded versus crushed) with dividers to avoid cross-contamination. When handling aggregates, drop heights should be minimized to also negate segregation and breakage. And, finally, when transporting aggregates, excessive vibration should be avoided to eliminate the possibility of segregation.

VI. Introduction to Cement

The most widely used type of cement is named after the limestone cliffs on the Isle of Portland in England. **Portland cement** acts like a type of glue that holds aggregate together to form concrete. The primary thing to remember about cement is that it hardens by **hydration**, which is the process of water chemically reacting with cement powder. This reaction is a tightly controlled process that is very sensitive to amount of water used, the ambient temperature, and a few other factors, to be discussed shortly.

Calling cement Portland cement actually means that it follows a specific creation process. Portland cement combines ground up lime-bearing (e.g., limestone, chalk) and clayey materials (e.g., slate, shale, clay). These initial compounds are placed in a kiln at 2,800°F, and walnut-sized chunks, caller clinkers, are produced. Lastly, clinker is ground up into a powder, and gypsum is added to avoid any hydration between the manufacturing plant and site mixing. A photo of a cement plant kiln is shown in **figure 6.10**.

When choosing the initial lime and clay components for Portland cement, the **chemical makeup** of

FIGURE 6.10   Cement kiln.

TABLE 6.5 MAIN COMPOUNDS OF PORTLAND CEMENT

Compound	Short-Hand Notation	Usual Range by Weight (percent)	Effect on Reaction	Benefits
Tricalcium silicate $3CaO \times SiO_2$	C_3S	45–60	Rapid	Leads to high, early strength Liberates heat quickly
Dicalcium silicate $2CaO \times SiO_2$	C_2S	15–30	Slow	Contributes to strength at time > 1 week Controls heat of reaction Leads to higher ultimate strength
Tricalcium aluminate $3CaO \times Al_2O_3$	C_3A	6–12	Depends on gypsum content	Minimal C_3A leads to sulfate resistance
Tetracalcium Aluminoferrite $4CaO \times Al_2O_3 \times Fe_2O_3$	C_4AF	6–8	None	Reduces energy needed to produce clinker
Gypsum	CSH	varies	Slow	Retards hydration

the materials can lead to vastly different results. The various chemical compounds in cement can control reaction speed, heat liberation, strength development, and sulfate resistance. **Table 6.5** lists the main compounds found in Portland cement and indicates the benefits of each compound pertaining to the hydration reaction.

For different applications, different cement compound mixes may be required. To help streamline design, ASTM has standardized a naming system that allows for uniform selection of compounds and subsequent performance throughout the industry. **Table 6.6** provides a list of the common cement types and indicates their uses.

In addition to chemistry, another factor to consider with cement is its **fineness**. The size of the particles in cement can directly affect hydration. Specifically, as fineness increases, we encounter a larger surface area per volume for particles, which leads to faster hydration. The appropriate term to describe the surface area per volume for a given particle is the **specific surface area** (SSA):

$$SSA = \frac{surface\ area}{volume} = \frac{SA}{V}.$$

Hydration is a chemical reaction where SSA and cement chemistry play an important role in controlling the rate of reaction. During this reaction, the cement and water mixture go

TABLE 6.6 CEMENT TYPES COMMONLY USED IN INDUSTRY

Cement Type	Use
I[1]	General purpose cement, when there are no extenuating conditions
II[2]	Aids in providing moderate resistance to sulfate attack
III	When high, early strength is required
IV[3]	When a low heat of hydration is desired (in massive structures)
V[4]	When high sulfate resistance is required
IA[4]	A type I cement containing an integral air-entraining agent
IIA[4]	A type II cement containing an integral air-entraining agent
IIIA[4]	A type III cement containing an integral air-entraining agent

Notes:

1. Cements that simultaneously meet requirements of type I and type II are also widely available.

2. Type II low alkali (total alkali as Na_2O < 0.6 percent) is often specified in regions where aggregates susceptible to alkali-silica reactivity are employed.

3. Type IV cements are only available on special request.

4. These cements are in limited production and not widely available.

through three stages. In the fluidlike first stage, chemicals on the surface of the cement particles react with the water, forming a gel. The presence of tricalcium aluminate (C_3A) liberates heat in this stage, while the presence of gypsum suppresses the reaction and ultimately controls the time to the **initial set**, which is where the mixture stiffens from a plastic to solid state. The calcium silicates (C_2S and C_3S) react with water to form calcium-silicate-hydrate (C-S-H). It is important to note that the more C_2S in a mix, the more C-S-H is formed—which is what gives concrete its ultimate strength. However, C_3S reacts with water more rapidly, contributing to a faster set and lower strength. In the setting stage of hydration, the hydrate crystals formed from the first phase begin to take over the space between particles. In the final solid stage of hydration, almost all water has reacted with the particles. In this stage, the hydration reaction process significantly slows down but continues to occur over a long period of time.

Water-to-Cement Ratio

While we have discussed how the cement particle's chemistry and size affect a concrete mixture's development, it should also be noted that water's availability and chemistry play an important role in this process. First, it is important to note the *ambient temperature* throughout this process, as it affects the availability of water for the reaction. High temperatures can lead to evaporation, while cold temperatures can lead to ice crystals forming or expansion cracks—both of which can impede the desired reaction and subsequent ultimate strength of your mix. Second, *how much water you add* with respect to cement affects the ultimate strength of the mix. This ratio is known as the water-cement (w/c) ratio and is given by

$$w/c = \frac{\text{weight of water in mix}}{\text{weight of cement in mix}}.$$

Cement typically needs 0.22–0.25 kilograms of water per kilogram of cement for proper hydration. However, more water is typically needed for the workability in the concrete, and this excess can lead to voids in the hydrated crystalline structure and thus a reduced strength. In fact, we can quantify this relationship between cement, water content (w/c), and compressive strength (f_c') using Abram's law:

$$f_c' = \frac{A}{B^{w/c}},$$

where A and B are constants for different cements.

Lastly, the chemistry of the water you had can have a great effect on your mix. While, technically, you can use any potable water in your mix, you need to be careful about impurities. Excessive chloride ions (added to drinking water to kill bacteria) can increase rebar corrosion, thus decreasing the tensile strength of a concrete member. Other impurities in water, such as sulfates, calcium, and other minerals, can also affect set time, strength, and durability as they interfere with the formation of C-S-H and other important hydration reactions.

VII. Concrete Admixtures

Admixtures are defined as any ingredients *besides* cement, water, and aggregate in concrete, and they can change the qualities of both the fresh (plastic) mix or hardened concrete. The addition of admixtures to your concrete design mix can help reduce construction costs through time savings, achieve certain concrete properties such as workability better than through additional water, ensure concrete quality under adverse weather conditions, and overcome certain emergencies during construction (delays).

Air entrainers, also called "foaming agents," produce microscopic bubbles in a concrete mixture. This additive usually comes in liquid form and is usually added at a concrete mix batch plant. The main role of this product is to provide freeze-thaw protection. It achieves this by allowing the concrete flexibility to expand into the air bubble voids instead of cracking during a freeze-thaw cycle. Air entrainers also allow for better workability of a mix and reduced segregation of aggregate particles, but they do lead to some reduction in strength.

Water reducers, commonly called "super plasticizers," reduce the amount of water required in a mix to maintain a certain workability. The result of using a water reducer will

EXAMPLE 6.7: FINDING ABRAM'S LAW CONSTANTS FOR CEMENT

For a cement mix, various compressive strengths were obtained by testing cement cylinders with different w/c ratios. When the data was plotted (*w/c* on the x-axis and $\log_{10}(f'_c)$ on y-axis), a regression analysis yielded the following results:

$$y = -1.3041x + 5.1144.$$

Determine the Abram's law constants A and B based on the regression analysis.

SOLUTION

For this scenario we know Abram's law,

$$f'_c = A \cdot B^{-w/c}.$$

Since we plotted $\log(f'_c)$ on the y-axis, we can use properties of logarithms to restate Abram's law in a form that matches our regression equation,

$$\log(f'_c) = \log(A) + \left(-w/c\right)\log(B),$$

or by rearranging,

$$\log(f'_c) = -\left(w/c\right)\log(B) + \log(A).$$

We also know that based on our original regression equation,

$$y = \log(f'_c).$$

Therefore,

$$\log(f'_c) = -1.3041x + 5.1144,$$

which has the same form as the rearranged Abram's law.

Taking the antilog,

$$f'_c = 10^{-1.3041x} + 10^{5.1144}.$$

Looking back at the rearranged form of Abram's law, we find,

$$A = 10^{5.1144} = 130,136.76,$$

$$B = 10^{1.3041} = 20.14.$$

yield high slumps that can be maintained for longer periods than a normal concrete. Water reducers are helpful in increasing workability while minimizing permeability (because of excess water pockets) and increasing strength.

Set retarders delay the setting of concrete, allowing for easier placing and finishing. They are mostly used in hot weather placement to provide time for special finishes (e.g., exposed aggregate). While they behave much like a cement with high C_2S values (hard to find), set retarders allow you to use a more common type of cement while achieving the results you want—saving money. Conversely, **set accelerators** increase the rate of setting, which is helpful in cold weather placement, repair work, setting anchor bolts, or anytime when high early strength is needed.

Not all additives to concrete are chemical. Minerals such as blast furnace slag (a byproduct from steel production) serve as filler and behave with similar properties as cement. **Pozzolans** (close to their cousin pozzolana, or volcanic ash) are another mineral example of a cement replacement. This category consists of industrial byproducts such as fly ash (a by-product of coal-burning power plants). These mineral additives combine with lime formed during cement hydration and have the same effect as adding C_2S (delayed strength buildup), which helps with increasing concrete workability.

VIII. Concrete Mixing

The combination of cement, aggregate, and water is known as concrete. When mixed appropriately, Portland cement concrete should achieve a desired strength, weather resistance, impermeability, and abrasion or wear resistance. Another advantage to concrete (over stone or other materials) is that it can be molded or shaped into anything you desire. The ingredients are readily available and inexpensive, and as we have learned, we can even recycle industrial waste. All of these benefits have a catch—how a concrete structure performs during its usable life depends on the original mix design, how it was transported, and how it was placed and cured.

Mixing and Handling Fresh Concrete

Ready-mixed concrete, commonly used for large construction jobs, is made at a central plant and delivered to a job site in trucks with rotating drums (see **figure 6.11**). Once on site, the concrete is either placed directly into formwork using a chute on the back of the truck or poured down the chute into a bin, where a pumping truck subsequently transfers the concrete into the desired location on site (see **figure 6.12**). Pumping trucks are especially useful for high-rise construction or areas with difficult access because of the flexibility and agility of the nozzle head.

After the concrete is placed, vibration is used to ensure that the formwork is completely filled and there are no large air pockets; at this point, potential problems can arise. If an improper mix was used, you may see segregation or bleeding. **Segregation** is the separation of coarse and fine particles in wet concrete and tends to occur when not enough fine sand is present. The result of segregation is

FIGURE 6.11   Ready-mix concrete plant and trucks.

easily viewable: large void pockets, called **honeycombing** (**figure 6.13**). When concrete is overly wet, has low cement content, has low fine sand content, or is over-vibrated or over-troweled, bleeding can occur. **Bleeding** is when water rises and sits on the concrete surface. The end result of the bleeding process is the migration of water with cement and fine aggregates to the surface of concrete and crack formation around the lower, larger aggregates. This process can lead to the loss of entrained air and scaling (peeling of the thin surface layer because of freeze thaw).

When handling concrete, we talk about its workability; **workability** is the ease with which concrete can be placed without segregation. The problem with this term is that there is no reasonable test for workability. Instead, we can test a concrete's consistency using the slump test, which involves forming a specimen of concrete inside a metal cone, then removing the cone and measuring how much the top of the concrete settles, or "slumps." The **slump test** is

FIGURE 6.12   Ready-mixed concrete being placed via a pump truck.

FIGURE 6.13   Example of honeycombing.

an index test that measures fluidity. The procedure for the slump test involves the following steps (see **figures 6.14a–6.14d**):

- Wet the inside of the metal cone and place it on a flat surface.
- Fill the cone in three layers, tamping each layer with a metal rod.
- Strike off the top so the concrete is exactly flush with the top of the cone.
- Lift the cone up and away.
- Measure the settlement or "slump" of concrete with respect to the height of the cone.

Typically, a slump value for a concrete is specified in job specifications. Slump increases with increasing w/c ratio, coarse aggregate content, or the quantity of air entrainer in the mix; conversely, slump decreases with increasing cement fineness and higher ambient temperatures.

Curing Concrete

Curing is the process of maintaining satisfactory concrete moisture content (MC) and temperature for a period of time with the purpose of ensuring ongoing hydration. Proper curing leads to increased strength, water tightness, freeze-thaw resistance, durability, abrasion resistance, and chemical resistance. Improper curing can lead to plastic **shrinkage** because of water evaporation. As shrinkage continues and rigidity increases, stresses build up and lead to the cracking of your concrete.

Figure 6.15 emphasizes the benefits of proper curing, depicting the effect of partial moist curing versus full moist curing. As expected, the compressive strength of the specimens is mostly set after seven days, with full strength reached by twenty-eight days. However, this figure also shows (via the solid line) that if a concrete specimen was moist cured continuously, its compressive strength after ninety days would be 123 percent of its twenty-eight-day strength! While this is exciting to think about in terms of underwater

FIGURE 6.14 Slump test procedure: (a) Slump cone; (b) tamping procedure; (c) removing the cone; (d) height measurement.

structures, for most sites, continuous curing would be expensive only to have a small return in strength for your effort.

There are several strategies to ensure proper curing. First, the presence of water should be maintained, usually through ponding, spraying, fogging, or wet coverings over the concrete. Second, moisture loss should be prevented through the use of impervious sheets, curing compounds, or leaving forms in place. Lastly, temperature should be monitored and kept constant; this can be achieved by steaming, insulating blankets, heaters, or through the use of heated aggregate.

Testing of Concrete

Because of concrete's brittle nature, we generally test concrete with a **compressive strength test**, which places a machine-controlled load on a concrete cylinder, which increases until failure (**figure 6.16**).

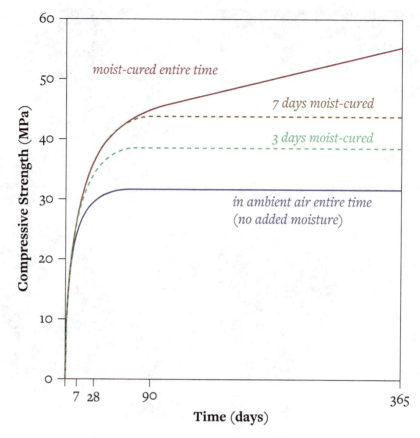

FIGURE 6.15 Effect of partial moist curing versus full moist curing of concrete.

The compressive strength (f_c') of a concrete is measured as the maximum load at failure (P_{max}) per the cross-sectional area (A_0) of specimen, or

$$f_c' = \frac{P_{max}}{A_0}.$$

In practice, these tests are used for quality-control purposes, acceptance of concrete, or estimating the concrete strength in a structure for the purpose of scheduling construction operations, such as form removal. A concrete's compressive strength is specified in job contract documents and strictly checked to ensure that design strength is met. To ensure that field and lab concrete mixes are comparable, ASTM procedures (C31 and C39) are used to reduce errors.

If we are looking to quantify the tensile strength of concrete, we cannot simply perform the same types of tests as we can with a ductile material, such as steel. Having a Poisson's ratio between 0.11 and 0.21, it is difficult to attain the tensile strength of concrete without a cracking brittle failure occurring. However, using indirect methods, we can measure the cracking to indicate tensile failure. The **split tension test** is such an indirect test that applies a compressive force to a concrete specimen in such a way that the specimen fails because of tensile stresses developed in the specimen (**figure 6.17**).

The tensile stress at which the failure occurs is termed the tensile strength (T) of concrete and is given by

$$T = \frac{2P_{max}}{\pi L d},$$

where

P_{max} = load at failure,

L = specimen length, and

d = specimen diameter.

In general, the value of the tensile strength varies from 360–450 psi, or approximately 10 percent of compressive strength.

FIGURE 6.16 Compressive strength test.

In highway or runway concrete pavements, it is often necessary to determine flexural strength. The **flexural strength test** measures a concrete's performance in bending (both compression and tension) via a three-point loading scenario (**figure 6.18** and described

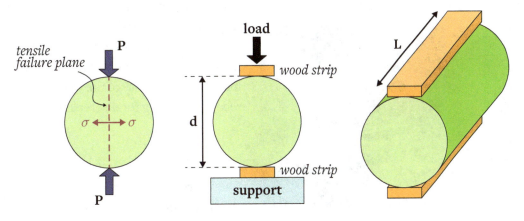

FIGURE 6.17 Split tension test.

FIGURE 6.18   Flexure strength test.

in chapter 2). The resulting value of the testing is the modulus of rupture (MOR). Flexural MOR is about 10 to 20 percent of compressive strength; the magnitude of the MOR depends on the type, size, and volume of coarse aggregate in the concrete mix. For a three-point loading scenario, we calculate the MOR through the maximum application of a load before failure, P_{max}, for a given specimen length, L, and cross-sectional dimensions of b (width) and d (height).

$$MOR = \frac{3P_{max}L}{2bd^2}$$

Lastly, there are several mechanical tests that can be performed on concrete already in place to determine strength and quality. One such test is the **Schmidt hammer test**, which measures hardness as correlated to strength via the rebound of a spring-loaded mass impacting against the surface of the sample (**figure 6.19**).

This test is nondestructive and simple to perform with the tool shown in figure 6.20. One destructive testing method is the **Windsor probe test**, which creates small holes in the concrete. This test measures penetration resistance by shooting probes into concrete. The penetration depth measured is inversely proportional to strength. Lastly, **ultrasound test-**

FIGURE 6.19   Schmidt hammer test.

ing is a nondestructive method used for detecting defects. Sound waves are sent into the concrete, and two transducers measure the velocity of the wave sent and received, looking for discrepancies.

IX. Concrete Alternatives

There are many alternatives to traditional concrete mixes, depending on the function of your project; everything from the aggregate to the admixtures, to the cement type can be altered. For example, to create a *lightweight concrete*, one must start with a lightweight aggregate, such as Styrofoam beads, volcanic pumice, slag, or another engineered product (**figure 6.20**). With a lighter mix, you may be able to get a compressive strength up to 2,500 psi, a sacrifice in compressive strength compared to normal weight concrete. Conversely, to develop a *heavyweight concrete*, you would employ heavy-use natural or steel aggregates. As mentioned previously, this type of mix is most appropriate for the isolation of radioactive sources.

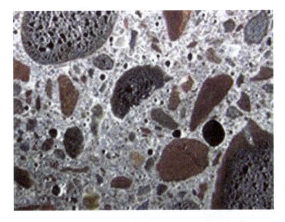

FIGURE 6.20   Lightweight concrete; note that the porous aggregate results in lower density.

If your job requires a *high-strength concrete* (greater than 6,000 psi), then your best bet would be to employ superplasticizers and a very low w/c ratio. One might find this type of concrete used in high-rise structures. Along the additives line, a *high workability concrete* (high slump) would also employ superplasticizers with normal w/c in order to work in closely bunched rebar, for rapid concrete placement in slabs, for underwater concrete placement, or for compact concrete surfaces.

Lastly fiber-reinforced concrete is another alternative to normal concrete. Fiber-reinforced concrete uses thin, neddle-like fibers to create a mattress effect in concrete, replacing rebar. This type of concrete holds up well after cracking and is good for earthquake zones or for thin placement applications (tunnel linings, pavements, floor slabs).

X. Introduction to Asphalt

The primary use of asphalt for civil engineers is for pavement construction and maintenance. Three of the most common types of asphalt used include asphalt cement, cutback asphalt, and emulsion asphalt. **Asphalt cement** is considered a hot-mix asphalt, requiring heat to allow the binder to flow for placement, and it is used in basic paving. Cutbacks and emulsion asphalts are cold-mix asphalts, which are of a lower quality than asphalt cement and best used for small repairs. **Cutback asphalt** is asphalt cement blended with solvent that cures by solvent evaporation. It is very fast to cure but generally expensive, highly polluting, and highly flammable. **Emulsion asphalt** is an asphalt cement that is suspended in water and emulsifying soap to keep the asphalt particles from adhering to each other.

After this asphalt mixture is laid, evaporation removes the water, and the asphalt behaves more like pure asphalt cement. In an emulsifying mix, it is important to understand if the emulsifying agent is anionic or cationic, because it will affect the asphalt's ability to adhere to certain aggregates. Anionic mixes will be attracted to cationic aggregate (e.g., limestone), whereas cationic mixes will be attracted to anionic aggregate (siliceous materials).

XI. Asphalt Properties

Absolute viscosity (sometimes known as dynamic viscosity, μ) quantifies the magnitude of a substance's internal friction and is measured by the shear stress (τ) per given strain rate ($\dot{\varepsilon}$) resisting a flow:

$$\mu = \frac{\tau}{\dot{\varepsilon}} = \frac{\text{Force} \cdot \text{Time}}{\text{Length}^2}.$$

The most common units used to measure viscosity are poise (P) (= 1 g/cm×s = 0.1 Pa×s) and lb-sec/ft² (= 478.8P). Viscosity is measured in the lab using a traditional **viscometer**, which is U-shaped, as depicted in **figure 6.21**.

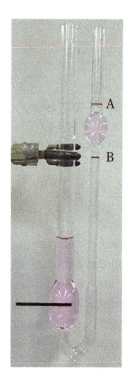

The substance to be measured is placed in the tube and placed into a bath to keep the temperature at 140°F. A vacuum is applied to raise the liquid to the upper mark in the tube. A timer is then started, and the time it takes the fluid to drop from the upper mark to the lower mark is then used with a calculation table supplied with the viscometer to give you your viscosity value.

Kinematic viscosity (ν) quantifies the ratio of dynamic viscosity (μ) of a fluid to the density (ρ) of that fluid:

$$\nu = \frac{\mu}{\rho} = \frac{\text{Length}^2}{\text{Time}}.$$

The most common units used in measuring kinematic viscosity are stoke (= 1 cm²/s) or ft²/s (= 929 stokes).

As indicated in **table 6.7,** the viscosity of asphalt has temperature susceptibility, meaning its value changes with temperature.

While the relationship of temperature and viscosity is actually highly nonlinear, if we plot log-log viscosity versus temperature, a linear scale actually emerges. This relationship is commonly referred to as the asphalt-viscosity-temperature susceptibility (**A-VTS**) **relationship**, and we use this to help predict asphalt performance.

FIGURE 6.21

Viscometer.

As emphasized in **figure 6.22**, the steeper the line, the greater the temperature susceptibility; thus, flat lines are optimal. We can use the knowledge gained from the A-VTS to help us in design. For example, in hot climates, we should select asphalt with a softer grade for the cold temperatures and, conversely, a harder grade for the warmer temperatures.

Chemical Properties

The properties of asphalt can be complex and highly dependent on the crude oil source and refining method. Ultimately, the final chemistry controls how asphalt will behave. Generally, asphalt is 80–87 percent carbon based, with 9–11 percent hydrogen, 2–8 percent oxygen, 0–1 percent nitrogen, 0.5–7 percent sulfur, and 0–0.5 percent trace metals. The most complex molecular chains, which include asphaltenes, resins, and oils, play the largest role in how asphalt behaves (see **table 6.8**).

TABLE 6.7 APPROXIMATE VISCOSITIES OF SELECTED LIQUIDS/SOLIDS

Liquid	Viscosity (centistokes)
Water (20°C)	1
Asphalt (104°F)	300
Honey	2,000
Molasses	5,000
Asphalt (natural)	10,000

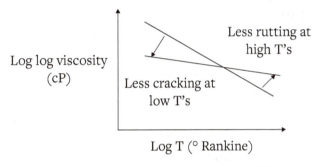

FIGURE 6.22   Asphalt-viscosity-temperature susceptibility

XII. Superpave and Performance-Based Binders

The **Superpave** (short for "superior performing asphalt pavements") system for asphalt grading and testing is a result of a $150 million government grant (Strategic Highway Research program) in the 1990s to improve the performance and durability of the US highway system.

The Superpave system ties asphalt cement and aggregate selection into the mix design process, with emphasis on projected traffic *loads* and *climate*. The results of this new mix design system were the minimization of

TABLE 6.8 ASPHALT INGREDIENTS AND THEIR ASSOCIATED BEHAVIORS

Component	Behavior
Asphaltenes	Viscosity and adhesion
	If absent, hard to compact asphalt pavement
Resins	Act as agents to disperse asphaltenes in oil
Oils	Medium for other chemicals

road deformation and cracking due to fatigue and temperature fluctuation. Superpave requires the characterization of binders (asphalt cement) through tests conducted at the maximum, minimum, and intermediate range of service temperatures projected for the location of the mix (intermediate is defined as the average of the minimum and maximum temperatures +4°C).

To begin characterizing a binder, the **rolling thin-film oven** (**RTFO**) is used to imitate short-term aging. Binder is put in bottles and rotated in a forced-draft oven at 163°C (325°F) for 75 minutes so that fresh asphalt is continuously exposed on the sides of bottles. The weight at the end of the test determines the mass lost and simulates the effect of heat and air on asphalt in the field during mixing and placement. Next, the **pressure aging vessel** (**PAV**) **test**, which imitates long-term oxidative aging of the binder, is performed. In this test, residue from the RTFO is put into the PAV at 2.1 MPa (305 psi) at a temperature of 90°C –110°C. Through this process, the binder ends up with five- to ten-year-old properties. After this, the flash-point test known as the **Cleveland open cup test** is performed to determine when asphalt ignites (safety test). To perform this, the binder is heated from underneath, and a flame is passed over the binder. When the sample vapors ignite momentarily in the air, the flash-point temperature is recorded.

To further characterize binder performance with respect to temperature, a **bending beam rheometer test** is implemented next. Binder is preprocessed in the RTFO and the PAV and then molded into a "beam." The beam is tested in bending, and its deflection is measured to find the rate at which the *binder relieves stress through plastic flow*. This is useful for understanding how binders will propagate flaws, such as cracks caused by thermal changes.

Classification of Asphalt

There are currently four methods for classifying asphalt: (1) based on additives (2) grading by penetration, (3) grading by viscosity, or (4) the Superpave system.

When **additives** are included in an asphalt mix to change the setting rates, we grade to denote this. **Cutback asphalts** are named with a prefix that indicates its cure rate followed by a number that represents the absolute viscosity of the asphalt at 140°F in poise. For example, RC-800 is a rapid curing asphalt with an absolute viscosity of 800 poise at 60°C (140°F).

Table 6.9 provides a summary of the cutback naming system. **Emulsion asphalts** are named with a prefix that includes the emulsion polarity (C for cationic, nothing for anionic) and its setting rate (rapid, medium,

TABLE 6.9 NAMING SYSTEM AND SOLVENT FOR CUTBACK ASPHALT

Prefix	Behavior	Solvent	Cure Time (hour)
RC	Rapid curing	Gas/naptha	4–8
MC	Medium curing	Kerosene	12–24
SC	Slow curing	Diesel fuel	48–60

From ASTM D2028, ASTM D2027, ASTM D2026

TABLE 6.10 NAMING SYSTEM FOR EMULSION ASPHALTS

Type	Bonds Best With	Prefix	Behavior	Viscosity Grades	Viscosity Range* (SFS)
Anion	Limestone dolomite	RS	Rapid setting	RS-1	20–100*
				RS-2	75–400
		MS	Medium setting	MS-1	20–100*
				MS-2	100+*
		SS	Slow setting	SS-1	20–100*
Cation	Granite sandstone	CRS	Rapid setting	CRS-1	20–100
				CRS-2	100–400
		CMS	Medium setting	CMS-2	50–450
		CSS	Slow setting	CSS-1	20–100*

* Denotes that the Saybolt viscosity test performed 25°C/77°F; all others at 50°C/122°F

Adapted from ASTM D977-17 and D2397

or slow setting) followed by a number (1 or 2) that represents increasing viscosity. For example, CRS-2 is a cationic, rapid-setting, high-viscosity asphalt, while MS-1 would be an anionic, medium-setting, low-viscosity asphalt. **Table 6.10** provides a summary of the naming system for emulsion asphalts.

Asphalt classified by **penetration grading** is given a grade that is measured by placing a needle with 100 grams of weight on the binder at 25°C (77°F) for 5 seconds. The penetration is recorded in 0.1-millimeter units and multiplied by ten. The grade is then determined using the ranges in ASTM D946 (a summary of the ranges is provided in **table 6.11**). For example, a penetration of 20.1 millimeters would be reported as 201 and be graded as 200–300.

Asphalt classified by **viscosity grading** can have a few naming systems depending on the type of asphalt. Asphalt cement in its original form (not heat tested in RTFO) is given the prefix of AC and followed by a number that represents the viscosity in poise at 60° C

TABLE 6.11 PENETRATION GRADES

Penetration Grade	Minimum Penetration (mm) at 25°C	Maximum Penetration (mm) at 25°C
40–50	40	50
60–70	60	70
85–100	85	100
120–150	120	150
200–300	200	300

Adapted from ASTM D946

TABLE 6.12 VISCOSITY GRADES AND ASSOCIATED PENETRATION FOR ORIGINAL ASPHALT
CEMENT

Viscosity Grade	Viscosity (P) at 60°C	Minimum Penetration (mm) at 25°C
AC-2.5	250 ± 50	200
AC-5	500 ± 100	120
AC-10	$1,000 \pm 200$	70
AC-20	$2,000 \pm 400$	40
AC-30	$3,000 \pm 600$	30
AC-40	$4,000 \pm 800$	20

Adapted from ASTM D3381

(140°F) divided by 100. For example, AC-20 is asphalt cement with a viscosity of 2,000 P at 140°F. **Table 6.12** provides a summary of values required for asphalts graded using viscosity tests as well as their associated penetration values.

Viscosity-graded asphalt cement can also be graded a second way, using *asphalt cement residue* left over from a RTFO test, which is used to simulate aging. If this method of grading is used, the grade is given the prefix AR and followed by a number that represents the viscosity in poise at 60° C (140°F). For example, AR-2000 is asphalt cement with a viscosity of 2,000 P at 60°C. **Table 6.13** provides a summary of values required for asphalt residues graded using viscosity tests as well as their associated penetration values.

As you may have noticed there are many discrepancies in the grading systems when it comes to penetration values. As shown by **figure 6.23**, there is overlap between viscosity and penetration grading! This can make things difficult in specifying the performance of an asphalt. Thus, to alleviate this confusion, the Superpave system was created.

As discussed earlier, the **Superpave** system classifies asphalt with both *load* and *climate* in mind. Different form the other classification systems described in this chapter,

TABLE 6.13 VISCOSITY GRADES AND ASSOCIATED PENETRATION FOR RTFO ASPHALT
CEMENT RESIDUE

Viscosity Grade	Viscosity (P) at 60°C	Minimum Penetration (mm) at 25°C
AR-1000	$1,000 \pm 250$	65
AR-2000	$2,000 \pm 500$	40
AR-4000	$4,000 \pm 100$	25
AR-8000	$8,000 \pm 2,000$	20
AR-16000	$16,000 \pm 4,000$	20

Adapted from ASTM D3381

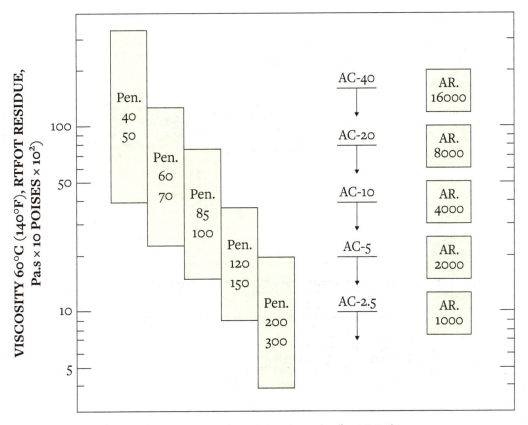

FIGURE 6.23   Comparison of penetration grades and viscosity grades (from FHWA).[2]

Superpave classification was designed and is controlled by the United States Department of Transportation's Federal Highway Administration (USDOT FHWA.) Asphalt classified by the Superpave system starts with the letters "PG" and is followed by the maximum and minimum design temperatures in °C. For example, PG 52–28 means a Superpave with a maximum service temperature of 52°C and minimum service temperature of –28°C. The test is performed at 10°C above the Superpave-rated temperature. (For example, a binder with a Superpave grading of PG 64–22 means that its minimum rated temperature is –22°C; thus, it is tested at –12°C.) Maps from the FHWA's Long-Term Pavement Performance (LTPP) program are shown in figures 6.24(a) and 6.24(b). These maps, created with the LTPP BIND program, show a 98 percent reliability in the projected high and low temperatures of pavement in a given location across the US.

---

2   Prithvi S. Kandhal and Rajib B. Mallick, "Pavement Recycling Guidelines for State and Local Governments," December 1997, https://www.fhwa.dot.gov/pavement/recycling/98042/07.cfm.

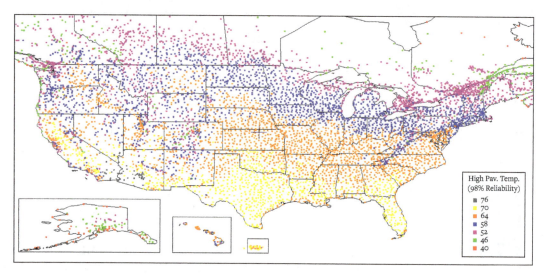

FIGURE 6.24A   Map of pavement high temperatures with 98 percent reliabilty, created using LTPP bind program.[3]

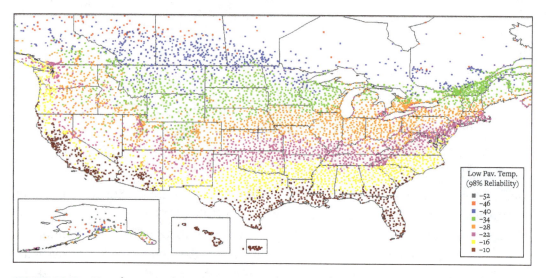

FIGURE 6.24B   Map of pavement low temperature with 98 percent reliability, created using LTPP bind program.[3]

XIII. Asphalt Concrete and Mix Design

The degree of flexibility of asphalt is proportional to the amount of asphalt cement in a mix. However, as asphalt cement increases, stability and skid resistance decreases. This is because particles begin to float instead of interlock, and the surface has little exposed aggregate.

---

3   "LTPP InfoPave," United States Department of Transportation, Federal Highway Administration, https://infopave.fhwa. dot.gov/Page/Index/LTPP_BIND.

Aggregate should be 70–75 percent by volume and 90–95 percent by weight in order to dominate the structural performance of the mix.

Hot-mixed asphalt that is used on road surfaces should provide stability (resistance to permanent deformation), resistance to fatigue cracking, resistance to thermal cracking, resistance to hardening or aging, resistance to moisture damage (stripping), resistance to skid, and workability during mixing, placing, and compaction.

To achieve the desired condition, there are several *aggregate gradations* we can consider. Open-graded aggregate (i.e., uniform graded) contains little or no fines and high void spaces. This type of grading will provide the most flexibility but will have lower stability and lower skid resistance. Dense-graded aggregate (i.e., well graded) provides an impermeable layer and good stability but may have reactivity with some asphalt binders.

To analyze the density and voids in a mix, the following equations can be used:

$$\text{VTM} = \text{voids in total mix} = \frac{V_v}{V_T}(100\%),$$

$$\text{VMA} = \text{voids in mineral aggregate} = \frac{V_v + V_b}{V_T}(100\%),$$

$$\text{VFA} = \text{voids filled with asphalt} = \frac{V_b}{V_b + V_v}(100\%),$$

where V_b is the volume of binder (asphalt cement), V_v is the volume of air/void space in the asphalt concrete, and V_T is the total volume of the asphalt concrete.

Just like concrete, additives can be added to asphalt to improve properties or give it special properties. **Fillers**, such as crushed fines (cement, lime, fly ash) can be added for better stability to improve binder-aggregate bonding or to fill voids (reduce binder required). **Extenders** are a chemical added to reduce binder amounts needed. Rubber can be added to improve mechanical properties and bonding as well. Lastly, antistripping agents, such as lime, are commonly added to ensure better bonding between the binder and aggregate.

Marshall Mix Design Test

The Marshall test is the most common test to determine stability and "flow" (deformability) in asphalt cement (see **figure 6.25**). To perform this test, complete the following:

- Compact asphalt concrete in mold that is 4 inches in diameter and 2.5 inches high by dropping a 10-pound hammer dropped from 18 inches above the specimen.
- Cure the specimen for 24 hours.
- Determine the density and voids.
- Heat to 140°F.

Marshall Test:

Asphalt specimen:

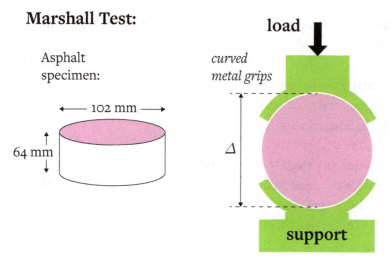

- Place on end in the "breaking head."
- Load at 2 inches per minute.
- Note that the max load in pounds equals "stability"
- Note that the deformation measured at max load equals flow, in number of 0.01 inches
- (e.g., 0.08 in. of deformation → flow = 8).
- Plot your results.

FIGURE 6.25   Marshall mix design test specimen and load configuration.

XIV. Concluding Remarks

In this chapter we discussed how aggregates are classified and sorted and how they behave. We looked at their application in cementitious materials and saw that they provide strength, at any size, to a mix. Understanding how to work with this naturally occurring resource is critical to all the subdisciplines in civil engineering; you cannot build something without encountering it. Earth in any form (rocks, sand, clay), while old, will forever be new as we engineer the world around us.

XV. Problems

1. You are preparing for a delivery of aggregate to be used in a concrete mix. You take a sample back to your lab and obtain the following values:

Aggregate Condition	Weight (kg)
Natural (wet)	23.8
Oven dry	20.3
Saturated surface dry	21.8

    a. What is the total moisture content of the natural aggregate?

    b. What is the absorption capacity of the aggregate?

c. Your concrete mix calls for 3,500 kilograms of this aggregate in the SSD condition. How many pounds of aggregate in the <u>natural condition</u> should be ordered?

2. Plot the following points to create a grain size distribution. Be sure to set your grain diameter to a log scale axis.

Sieve Size		% Passing		
ASTM	(mm)	Soil #1	Soil #2	Soil #3
No. 4	4.750	96	75	80
No. 8	2.360	94	74	78
No. 20	0.850	60	68	55
No. 30	0.600	48	20	53
No. 50	0.300	32	13	50
No. 100	0.150	25	13	44
No. 200	0.075	20	13	12

a. Using the plot, determine if the soil is uniform, well, or gap graded.
b. Use log interpolation to find the values for the #16 sieve (1.180 millimeters).
c. Determine the fineness modulus (FM) for the three soils.

3. The table that follows shows the grain size distribution for two aggregates and the specification limits for a compacted fill soil.

Sieve Size (mm)	19	12.5	9.5	4.75	2.36	1.18	0.600	0.300	0.150	0.075
Specification limits	90-100	80-85	70-80	60-70	50-55	40-45	30-35	20-25	n/a	n/a
Aggregate A percent passing	100	86	80	70	48	36	28	20	18	12
Aggregate B percent passing	88	80	62	60	56	48	36	25	14	2

a. Determine the **exact** values for the pinch points.
b. Given aggregate A costs $5,000 per ton, and aggregate B costs $1,000 per ton, if 20 tons of the mix is required, what is the minimum cost of the aggregate?

4. Explain how the following compounds affect the character of cement:
   a. Tricalcium silicate
   b. Dicalcium silicate
   c. Tricalcium aluminate
   d. Tetracalcium aluminoferrite

5. Specific surface area: Determine the specific surface area of the following objects:

    a. A montmorillonite clay particle 100 nanometers in diameter, 1 nanometer thick

    b. A 1-inch steel rod, 6 inches long

    c. A 5-inch diameter orange cut into four slices

6. Plot the following points and determine the Abram's law constants using a best-fit line. Compare the results obtained for w/c = 0.6 from the table and using your regression equation.

Water-to-Cement Ratio (w/c)	f'_c (MPa)
0.4	39.20
0.5	30.32
0.6	22.68
0.7	17.48
0.8	11.82

7. What would you add to your concrete mix to achieve the following qualities?

    a. Early mix setting

    b. Delayed mix setting

    c. Reduced curing rate in hot weather conditions

    d. Enhanced curing rate in cold weather conditions

    e. Enhanced workability

    f. Reduced water requirement

    g. Reduced segregation

8. A split tension test is to be performed on a concrete specimen with a diameter of 8 inches and length of 12 inches. What load will be required to fail this specimen if a conventional compression test determined the compressive strength to be 6,000 psi?

9. Grade the following asphalts:

    a. Kerosene based asphalt with an absolute viscosity of 25,000 centistokes at 60°C

    b. An asphalt to be mixed with limestone with a viscosity of 80 SFS at 50°C

    c. An asphalt cement residue with a penetration 45millimeters at 25°C and a viscosity of 1,800 P at 60°C

    d. An asphalt to be used on a federal highway in Michigan. Assume you can use 98 percent reliability figures from LTPP.

10. A pothole 8 inches deep, 2 feet long, and 1 foot wide will be filled with asphalt cement. The mix specified is provide in the table. Determine the weight of the individual components to measure in the field to fill the hole.

	Effective Specific Gravity	Percent by Volume
Asphalt cement	1	12.4
Coarse aggregate	2.7	57
Fine aggregate	2.65	20.6
Mineral filler	2.6	4
Air	-	6

7

Wood—A Renewable Material

I. Introduction to Wood

Wood was certainly among the first engineering materials used by humans—it is readily available without mining, processing, or alteration; it is easy to dimension and shape as needed; and it has a relatively high strength-to-weight ratio. It is a highly unusual material because of its different properties in different directions (**anisotropy**), which we usually associate with the wood grain, and because its engineering properties are strongly dependent on the degree of moisture in the wood. Wood is considered to be one of the "greenest" materials used in civil engineering since the source material is natural, renewable, and reusable; it is produced with relatively low energy input (often relatively close to the point of use); and it possesses thermal and other properties that reduce energy consumption during use.

While we rely on wood for a number of civil engineering applications, it also has many potential pitfalls, including natural defects that may affect engineering properties; the ability to absorb water that may lead to rotting, mold growth, and dimensional changes (swelling); and non-engineering aspects, such as flammability and susceptibility to attack by insects and other organisms.

II. Biological Composition and Structure of Wood

Unlike any other material covered in this book, wood has a biological origin that makes its composition unique and complex. That biology is responsible for wood's physical structure at the cellular level on up to the macroscale. With over 30,000 species of trees, there is a high variability in the engineering properties of wood. To help us identify these properties, we divide trees into two basic categories of *hardwoods* and *softwoods*.

The **softwoods** are identified by their needlelike leaves (for example, pine needles), and generally remain "leafed" all year round (although losing some portion of those leaves during the year) so that they are often referred to as *evergreens*. As cataloged by the US Department of Agriculture,[1] there are some 52 types of softwoods in North America alone, and about 500 worldwide, giving us some idea of the biological variety within this classification. You may also know that most softwoods produce cones, inside of which the seeds are produced, from which their other name is derived: *conifers*. Some examples of softwoods found in the Northern Hemisphere include spruce, larch, fir, hemlock, redwood, yew, cypress, and cedar. Of the 100 species in the United States that reach tree size and are commercially important for wood production, only 35 are softwoods.

Hardwoods are broad-leafed trees, and in temperate climates, those leaves change color in the fall and typically drop from the trees. Their seeds are contained in pods or fruit that is produced by the tree, thus forming the covering that we mentioned earlier. Perhaps the most common "fruit" that we are familiar with in North America is the acorn produced by oak trees. Again, the US Department of Agriculture[2] has a publication dedicated to cataloging the hardwoods of North America, in this case, 53 species, but there are literally thousands of hardwood species worldwide. Some examples of hardwoods include palm and yucca trees, oak, ash, elm, maple, birch, beech, cottonwood, and aspen.

Regardless of species, living wood cells have three functions: *structural support* of the tree, *transport* of nutrients, and *storage* of organic substances produced by the tree for food. Since trees used for wood in engineering are all **exogenous**—trees that add new growth by the development of growth rings of cells on the *outside* of the older wood—the only living cells in the tree are on the *outer* part of the tree! To understand this, let's look at the parts of a tree.

Working from the outside of a tree inward, the parts consist of the outer bark, inner bark, cambium, sapwood, heartwood and pith (**figure 7.1**). The **outer bark** is a dense, rough protective layer that was produced as part of this annual growth, but whose cells soon died. The **inner bark** consists of cells that are still living and provide downward transport of manufactured sugars for tree nutrients.

Inside the inner bark is the **cambium**, which is the source of all tree growth. The cambium begins each growth cycle as a single layer of cells that remains dormant during the winter but then expands to about eight to ten cells thick in the radial direction (i.e., outward from the center of the tree). As time progresses, further radial subdivision occurs, with some cells remaining as cambial cells, some to join the internal wood mass, and others to become part of the bark.

1   Harry A. Alden, "Softwoods of North America," US Department of Agriculture Forestry Service, September 1997, https://www.fpl.fs.fed.us/documnts/fplgtr/fplgtr102.pdf.

2   Harry A. Alden, "Hardwoods of North America," US Department of Agriculture Forestry Service, September 1995, https://www.fpl.fs.fed.us/documnts/fplgtr/fplgtr83.pdf.

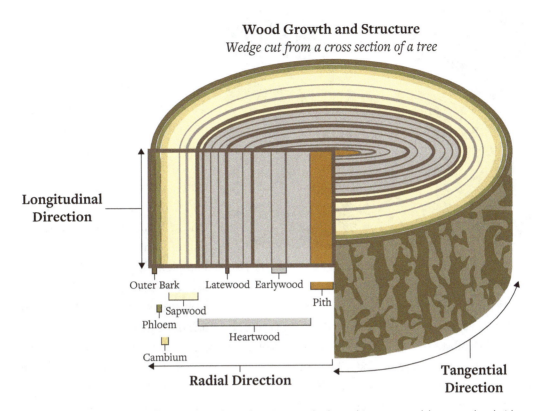

Wood Growth and Structure
Wedge cut from a cross section of a tree

Longitudinal Direction

Outer Bark   Latewood   Earlywood

Sapwood

Phloem

Heartwood

Pith

Cambium

Radial Direction

Tangential Direction

FIGURE 7.1   Parts of a tree from the outside working in: outer bark, cambium, sapwood, heartwood and pith.

The cambium surrounds the newest, still living wood cells that are collectively referred to as the **sapwood**. The most recently added wood in the sapwood region is further subdivided into latewood (develops later in the growing season, so as noted earlier, it is sometimes called summer or autumn wood) and earlywood (also called springwood). While a majority of the sapwood cells are dead, a small percentage will carry on metabolic processes. From an engineering point of view, this attribute is important since it results in a relatively high moisture content (MC) in the sapwood compared to the next innermost set of layers, known as the **heartwood**. Very few cells remain alive in the heartwood, as those cells have less and less access to water, and chemicals accumulate in the cell walls that inhibit biologic activity. The heartwood section of the tree is visually distinguished from the sapwood by its darker color (see **figure 7.2**), the result of the chemical changes that occur as it transitions from sapwood. The reduced moisture content in heartwood generally makes it the *more desirable part of the tree* from an engineering point of view. Finally, at the center of the tree lies the **pith**, which consists of parenchyma cells that subdivide for the height growth of the tree.

FIGURE 7.2   Maple tree showing contrast between heartwood and sapwood.

III. Chemical Composition of Wood

At its most basic level, wood solids are composed of about 50 percent carbon, 44 percent oxygen, and an average of 0.2–0.3 percent inorganic, mineral components consisting of potassium, magnesium, manganese and/or silicon, and 0.1 percent or less nitrogen. These basic building blocks in wood combine into much more complex, organic molecular constructs collectively known as **polymers**. While polymers are commonly associated with plastics, the term is actually much more generic in its application; in the case of wood, it describes the relatively long chains of molecular units that ultimately form the cell walls of the wood. The three types of polymers in wood are **cellulose** (40–44 percent by dry weight of wood), **hemicellulose** (15–35 percent), and **lignin** (18–35 percent), and together these are responsible for the mechanical properties of the wood. Cellulose and hemicellulose form the tree's cell walls, which in turn are known to us collectively as wood fibers. Lignin acts as a "glue" to hold cells together and gives the wood cells rigidity.

IV. Physical Properties of Wood

Anisotropic Nature of Wood

Perhaps the most important and unique aspect of wood as an engineering material is its **anisotropy** (*an-* meaning not, *iso-* for equal, and *-tropic* for affecting). This means that the wood will have *different properties in different directions* with respect to its cross-section. **Figure 7.3** shows how the three directions, **longitudinal**, **radial**, and **tangential**, are defined along the tree sections.

The **longitudinal direction** is so named because the wood cells generally align their long axes to this direction, which runs up and down the long axis of the tree or limb, with a resulting geometry that is much like a bundle of straws glued together. This direction will be very important for our engineering purposes since, among the three directions we're describing, it is by far the strongest and stiffest (highest modulus) and experiences the least geometric change with change in moisture content (shrinkage and swelling).

The **radial direction** runs from the pith of wood outward toward the bark, thus cutting across the growth rings and generally perpendicular to the cell walls. If we were, for example, compressing the wood in the radial direction, to use our bundle of straws analogy, we would be squeezing the straws laterally on their thin walls—you can imagine that the strength in this direction is much less than that along the longitudinal axis.

The third direction is **tangential**, and as the name implies; this is the axis that

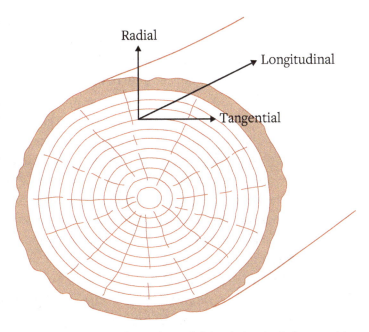

FIGURE 7.3   Three orthogonal axes; defining the longitudinal, tangential and radial directions for wood based on the growth rings and grain.

is perpendicular to the radial direction but tangent to any growth. It is important to note that in practice, it is certainly impractical to cut a tree so that all three axes are perfectly aligned with their axes, primarily because of the circular (i.e., nonlinear) growth rings (**figure 7.3**). Thus, when assigned axes to a piece of cut lumber, it will be according to the predominant axis for a particular dimension. As noted, because of the significant differences in properties among the three axes, we will pay careful attention to these differences when each property is covered in more detail.

Moisture Content and Its Effects

Other than its anisotropic nature, there is no other factor that impacts wood behavior as much as the **moisture content** (**MC**). The amount of moisture in wood will strongly affect its density or unit weight, strength, stiffness, swelling/shrinkage potential, susceptibility to attack by insects, potential for mold growth, and so on. The basic equation for MC is

$$MC = \frac{M_w}{M_o} \times 100\%,$$

where M_w = mass of water in the wood, and M_o = oven-dry mass of the wood (i.e., the mass of wood solids alone).

Wood is commonly found in three moisture conditions: green, kiln dry, and oven dry. Wood that has just been cut is known as **green wood.** Green wood has moisture absorbed in the wood's cell walls (**bound water**) and cell cavities (**free water**). While the idea of the two types of water in wood, bound and free, is conceptually appealing, the reality is more complex: There is a gradual transition between bound and free water, and there will be parts of the wood with different proportions of each. Some cell walls may have started to dry out while their cell cavities still contain free water.

Kiln drying and oven drying both involve the use of heat to reduce the moisture content of wood. The difference between these two processes is the temperature of the drying process and the desired end product. **Kiln drying** is performed at 155°F–180°F and aims to eliminate most of the free water, leading to better structural properties. The initial MC of a wood before kiln drying can be highly variable, as it depends on from where the wood was cut. For example, heartwood can start with a MC of 25 percent, while sapwood can start with a MC of 250 percent! Thus, before using wood as a structural material, it is usually kiln dried to *restrict future moisture-related swelling, limit fungus growth, and enhance mechanical properties.* After kiln drying wood, a typical MC is 12 percent.

Oven drying is performed at 220°F to eliminate almost all water, leaving the wood very brittle. Therefore, oven drying is practiced mostly to collect laboratory data.

Density and Specific Gravity

The density (mass per volume) and corresponding unit weight (force per volume) of wood varies widely according to tree species and moisture condition. The issue of moisture condition and its effect on density is further complicated by the fact that both the mass and volume are affected by changes in moisture (compare this to brick and stone, which change mass but not volume when absorbing moisture!) For wood, **density** can best be defined by one of the following equations:

$$\text{Solids density, } \rho_s = \frac{\text{mass of wood solids (oven dry)}}{\text{volume of wood solids ONLY}} = \frac{M_o}{V_s},$$

$$\text{Bulk density, } \rho_b = \frac{\text{mass of wood solids (oven dry)}}{\text{green volume of wood}} = \frac{M_o}{V_G},$$

$$\text{Oven dry density, } \rho_o = \frac{\text{mass of wood solids (oven dry)}}{\text{volume of wood (oven dry)}} = \frac{M_o}{V_o},$$

$$\text{Density at x\% moisture, } \rho_x = \frac{\text{mass of wood at x\% moisture content}}{\text{volume of wood at x\% moisture content}} = \frac{M_x}{V_x}.$$

Note that the same equations can be used to determine the unit weight of wood by simply replacing the mass with weight. Of these four definitions for density, the standard reported density (or unit weight) of the wood solids in the US is ρ_b (or γ_b), as this describes the amount of solid material in the initial (green) volume of the wood.

The density at a particular MC, ρ_x, is often used for structural calculations (since the mass or weight varies significantly with changes in MC) and weights used for shipping and crane lifts. As we will see in the next section, a common MC for wood is 12 percent after kiln drying so that one will often see ρ_{12} specifically referred to for a given type of wood.

The **specific gravity** definitions for wood all rely on the oven-dry mass as the basis, and it is only the volume that varies across the definitions, as follows:

$$G_s = \frac{\text{density of wood solids @ temperature, T}}{\text{density of water @ temperature, T}} = \frac{\rho_s}{\rho_w},$$

$$G_b = \frac{\text{bulk density of wood @ temperature, T}}{\text{density of water @ temperature, T}} = \frac{\rho_b}{\rho_w},$$

$$G_o = \frac{\text{oven dry density of wood @ temperature, T}}{\text{density of water @ temperature, T}} = \frac{\rho_o}{\rho_w},$$

$$G_x = \frac{\text{density of oven dry wood at moist volume @ temperature, T}}{\text{density of water @ temperature, T}} = \frac{\frac{M_o}{V_x}}{\rho_w},$$

where ρ_w is the density of water. Note that wood in the "green" state corresponds to that just after cutting, without significant drying. **Table 7.1** provides a summary of symbols used in the various density and specific gravity equations.

There are a number of methods for converting from the specific gravity at one MC to that at another, and for converting from one specific gravity definition to another (say, from G_b to G_x). These almost all involve measuring the degree of **shrinkage** in the material (change in dimensions or volume because of moisture loss/gain), which one would expect since both the change in mass and the change in volume need to be known to convert. Perhaps the simplest is the equation to convert G_b to G_x:

$$G_x = \frac{G_b}{1 - \frac{S_x}{100}},$$

TABLE 7.1 SUMMARY OF DENSITY AND SPECIFIC GRAVITY EQUATIONS

	Symbol	Mass Basis	Volume Basis
Density	ρ_s	OD	Solids only
	ρ_b	OD	Green
	ρ_o	OD	OD
	ρ_x	x% MC	x% MC
Specific gravity	G_s	OD	Solids only
	G_b	OD	Green
	G_o	OD	OD
	G_x	OD	x% MC

MC = moisture content
OD = oven dry

where S_x is the percent volumetric shrinkage given by

$$S_x = \left(\frac{\Delta V}{V_G}\right) * 100\%.$$

ΔV is the change in volume from the green condition to the MC x. Typical values of shrinkages range from 7–21 percent, but these values will be less if only kiln drying.

It should also be noted that because wood is anisotropic, there will be different rates of shrinkage along the three axes. **Table 7.2** compares typical shrinkage values along the axes, indicating that shrinkages can be greatest along the tangential axis.

There exists a large range in these values to accommodate for the fact that where the wood was cut from the tree changes the MC. This means that wood cut from the younger outer rings of the tree (sapwood) will shrink more than older inner wood (heartwood). When cut wood shrinks at uneven rates or magnitudes, this is known as **warping**. Warping is highly dependent on how wood is cut from the tree. For example, radial cuts, such as rift cuts (shown in **figure 7.4**), tend to warp the less-than-tangential cuts, such as slash cuts.

TABLE 7.2 SHRINKAGE ALONG GRAIN AXES

Axis	Typical Shrinkage
Axial/longitudinal	0.1–0.2%
Radial	2–8%
Tangential	4–14%

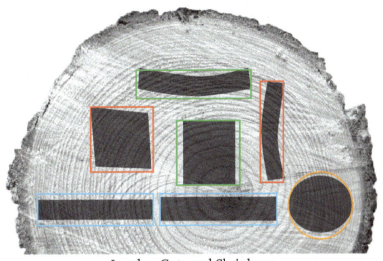

Lumber Cuts and Shrinkage

Legend: Typical Cuts

| — Round | — Flat Sawn | — Rift Sawn | — Quarter Sawn |

FIGURE 7.4   Common cuts and typical shrinkage shapes.

EXAMPLE 7.1: VOLUME AND WEIGHT PROPERTIES OF WOOD

A green piece of wood with a MC of 32 percent is *oven* dried to the dimensions shown in **figure 7.5**. The oven dried specific gravity is 0.60, and the volumetric shrinkage was measured at 12.6 percent.

    a. *Determine the weight of the wood solids.*

    b. *Determine the original, green volume of the wood.*

    c. *What should be the reported unit weight of the wood solids?*

    d. *How much did the wood weigh green?*

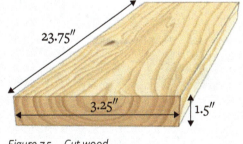

Figure 7.5   Cut wood.

SOLUTION

    a. Adjusting for weight units:

$$G_o = \frac{\gamma_o}{\gamma_w} = \frac{\dfrac{W_o}{V_o}}{\gamma_w},$$

we can rewrite the equation as

$$W_o = G_o \cdot \gamma_w \cdot V_o = (0.60)\left(62.4\frac{lb}{ft^3}\right)\left[(1.5\,in)(3.5\,in)(23.75\,in)\right]\left(\frac{ft}{12in}\right)^3 = 2.5\,1lb.$$

b. Rewriting the equation

$$S_x = \left(\frac{\Delta V}{V_G}\right) * 100\%,$$

we can find the original (green) volume:

$$V_G = \frac{V_o}{1-S_x} = \frac{\left[(1.5\,in)(3.5\,in)(23.75\,in)\right]\left(\frac{ft}{12in}\right)^3}{1-0.126} = 0.0767\,ft^3.$$

c. Calculating the unit weight,

$$\gamma_s = \frac{W_o}{V_G} = \frac{2.51\,lb}{0.0767\,ft^3}.$$

d. Rearranging the equation

$$MC = \frac{W_w}{W_o}(100\%)$$

we can solve for green weight:

$$W_G = (1+MC)(W_s) = (1+0.32)(2.51\,lb) = 3.31\,lb.$$

V. Mechanical Properties

Now that we have an understanding of the physical properties of wood, these will be critical to our study of wood mechanical properties. Beside tree species, the two most important determining factors for mechanical properties are MC and direction of loading (relative to the grain orientation). **Figure 7.6** illustrates the various loading scenarios on cut wood that will be discussed.

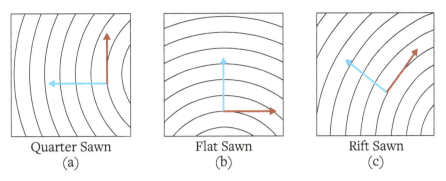

FIGURE 7.6   End grain appearance for (a) quarter sawn, (b) flat sawn, and (c) rift sawn lumber, showing the radial and tangential directions for loading.

Modulus of Elasticity

Due to the orthotropic nature of wood, it can be expected that the **modulus of elasticity**, or Young's modulus, denoted by E, will vary depending on the direction of loading. Despite many factors that cause potentially significant variations in the modulus values, the relationship among the moduli is relatively consistent. Generally,

$$E_L = 1 \text{ to } 2 \times 10^6 \text{ psi (longitudinal)},$$
$$E_R = 10\%(E_L) \text{ (radial), and}$$
$$E_T = 5\%(E_L) \text{ (tangential)},$$

where E_L, E_R, and E_T are the moduli in the longitudinal, radial, and tangential directions, respectively. As the earlier ratios indicate, the modulus of elasticity decreases with increasing MC. For every 1 percent increase in MC, there is a 1–3 percent decrease in the modulus of elasticity.

Tensile Strength

Tensile strength of wood is most commonly measured as the resistance to splitting perpendicular to the grain. There are few data on the resistance to tensile loading parallel to the grain. This is unfortunate, since wood used for structural purposes typically relies on the latter type of loading, particularly when the wood is used to resist bending forces (recall that bending stress is really a combination of compressive and tensile stresses). A major difficulty associated with testing wood in pure tension, as we do with metals and plastics, is that the wood often *fails first at the connection points used to grip the wood* rather than in the middle of the specimen where the true tensile stresses exist.

For this reason, modulus of rupture (MOR) data is often used as a conservative substitute for tensile strength—by conservative, we mean that it provides an estimate of tensile strength that is lower than the true value. This is because when tested in bending, the wood fails *first in compression* before the full tensile strength can be mobilized.

Nonetheless, there are tensile strength data for wood taken from a relatively limited number of tree species; **table 7.3** provides these tensile strength values for wood tested when green. To adjust these strengths for kiln-dry states (12 percent MC), hardwood strengths should be increased by about 32 percent and those of softwoods by about 13 percent.

TABLE 7.3 SELECTED EXAMPLES OF WOOD TENSILE STRENGTH PARALLEL TO GRAIN[3]

Hardwood Species	Tensile Strength		Softwood Species	Tensile Strength	
	kPa	lb/in²		kPa	lb/in²
Beech, American	86,200	12,500	Cedar, Western Red	45,500	6,600
Elm, cedar	120,700	17,500	Douglas Fir	107,600	15,600
Maple, sugar	108,200	15,700	Fir, California Red	77,900	11,300
Oak, pine	77,900	11,300	Pine, Eastern White	73,100	10,600
Poplar, balsam	51,000	7,400	Redwood, Virgin	64,800	9,400
Willow, black	73,100	10,600	Spruce, Engelmann	84,800	12,300

ᵃ Results on clear, straight-grained specimens tested green. To adjust these strengths for kiln-dry states (12 percent MC), hardwood strengths should be increased by about 32 percent and those of softwoods by about 13 percent.

Compressive Strength

One of the most important mechanical properties for wood is its **compressive strength**. Aside from bending, compression is one of the most common loading modes for structural applications. When wood is tested in compression, there are a number of potential failure modes, primarily because of the wood grain. Common failure modes are shown in **figure 7.7**.

Typical compressive strengths parallel to the grain range from 3,500 to 8,500 psi. When a specimen is loaded perpendicular to the grain, the anisotropy of wood's strength is apparent, with strength values usually around 15 percent of those from parallel loading. This is a reflection of the tubular structure of wood: In compression loading parallel to the grain, the wood tubes are loaded as columns, but adjacent tubes provide intermediate support to one another, preventing buckling from occurring. In this way, as we've noted before, wood is the original composite material! However, when loaded perpendicular to the grain, tube crushing is occurring relatively easily, resisted only by the rigidity of the cell walls.

Bending (or Flexural) Strength

The **bending strength** of wood is typically determined using the *four-point* bending device, as shown in **figure 7.8**. The four-point bending device applies load to the specimen at four

---

3   D. W. Green, J. E. Winandy and D. E. Kretschmann, "Mechanical Properties of Wood," in *Wood Handbook* (Madison, WI: United States Department of Agriculture, Forest Products Laboratory, 1999).

Most Common Failure Modes for Wood

Crushing   Shearing   Splitting   End rolling

FIGURE 7.7   Common failure modes of wood in compression.

points along its length (at each end, and at the third points of the span), hence the name for the device and test.

Just to confuse things, this test is sometimes called the *third-point* bending test because of the intervals across the loading span where the load-bearing blocks contact the beam. Recall from **chapter 2** that a four-point test is used for materials that may not be uniform or homogeneous in

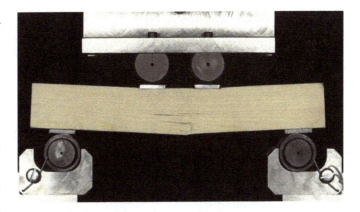

FIGURE 7.8   Wood specimen in a four-point bending device for determining bending strength.

composition. As described in **chapter 2,** between the two loading points on top of the specimen, the applied bending moment is theoretically uniform (constant) so that the measured bending strength should ideally reflect "average" behavior between those two points. In the case where the tested material may contain nonuniformities or defects, such as those found in wood, this span of uniform moment provides a means for measuring behavior that is more representative of the entire piece of wood, rather than that at an isolated location.

The maximum load carrying capacity of a material in bending is referred to as the modulus of rupture (MOR) and is proportional to the maximum moment that can be supported by the material (see **chapter 2**). Finding the MOR is complicated with wood since it is such an anisotropic material—the compressive strength being mobilized on the top of the bending test specimen is generally much lower than the tensile strength on the bottom of the specimen. As a result, the portion of the cross-section in compression fails before the tension zone, and the effective centroid of the bending shape begins to shift downward. Typical values for MOR of wood usually range from 6,000 to 15,000 psi.

Shear Strength

Shear strength parallel to the grain can be visualized using the "bundle of straws" model. If the bundle were gripped on each side and sheared parallel to the "grain" (lengthwise dimension of the straws), the straws would slide lengthwise across one another, distorting the straw stack (**figure 7.9a**). Conversely, if the stack were loaded in shear perpendicular to the grain, then the fibers would be able to support one another more effectively, leading to stronger resistance (**figure 7.9b**).

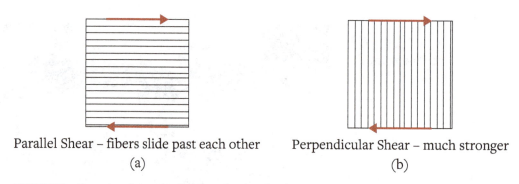

Parallel Shear – fibers slide past each other
(a)

Perpendicular Shear – much stronger
(b)

FIGURE 7.9   Examples of shear loading scenarios, bundle of straws model: (a) Loading parallel to the wood grain; (b) loading perpendicular to the wood grain.

VI. Defects in Wood Affecting Properties

Wood has a number of natural characteristics that can seriously affect the engineering properties of the material. With the possible exception of natural rock, no other engineering material has macroscale defects like those observed in wood that are considered acceptable

EXAMPLE 7.2: LOADING SCENARIOS ON WOOD

A piece of wood has the dimensions shown in figure 7.10a. The wood has the following properties:

Direction	Compressive Strength (psi)	Tensile Strength (psi)	Modulus of Elasticity (psi)
Longitudinal	5,450	15,475	1.4×10^6
Radial	725	1,410	1.28×10^5
Tangential	840	1,445	7.5×10^4

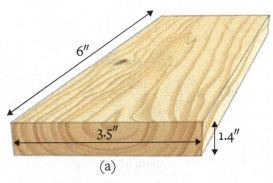

6"

3.5"

1.4"

(a)

Figure 7.10a

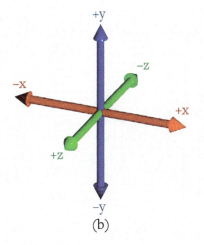

+y

−z

−x

+x

+z

−y

(b)

Figure 7.10b

a. *If the wood is loaded in compression in the x direction (see figure 7.10b, what will be the load in pounds at failure?*

b. *When wood is to be used as a column, it will be loaded in compression parallel to the grain. If deformation needs to be limited to 0.02 inches, what is the maximum load that can be applied?*

c. *If this piece is to be loaded like a joist in bending (i.e., load applied to the narrow edge, with the long dimension as the height), what is the maximum moment that can be supported without failure in either tension or compression?*

SOLUTION

a. The x direction in this scenario is tangential, so looking at the table, we will use $\sigma_c = 840$ psi.

$$P = \sigma_c \cdot A = (840 \text{ psi})(1.4 \text{ in})(6 \text{ in}) = 7,056 \text{ lb}$$

b. Parallel to the grain intimates a longitudinal direction; thus, we will use $E_L = 1.4 \times 10^6$ psi.

First, we calculate the allowable strain in the longitudinal direction (along the z axis):

$$\varepsilon_z = \frac{0.02 \text{ in}}{6 \text{ in}} = 0.033.$$

Next, we find the stress on the cross-section:

$$\sigma_z = E \cdot \varepsilon_z = (1.4 \times 10^6 \text{ psi})(0.033) = 4{,}667 \text{ psi}.$$

Then we can solve for the maximum load that we can apply:

$$P = \sigma_z \cdot A = (4{,}667 \text{ psi})(1.4 \text{ in})(3.5 \text{ in}) = 22{,}867 \text{ lb}.$$

A bending moment on this wood in a joist orientation implies that a force was applied in x direction, creating a combination of compression and tension in the longitudinal direction. Examining the cross-sectional area of the beam we find

$$I = \frac{bd^3}{12} = \frac{(1.4 \text{ in})(3.5 \text{ in})^3}{12} = 5 \text{ in}^4 \quad \text{and} \quad c = \frac{d}{2} = \frac{3.5 \text{ in}}{2} = 1.75 \text{ in}.$$

Thus, the maximum moment the beam is

$$M = \frac{\sigma_c \cdot I}{c} = \frac{(5{,}450 \text{ psi})(5 \text{ in}^4)}{1.75 \text{ in}} = 15{,}576 \text{ lb} - \text{in}.$$

(but still undesirable). As biological organisms growing in nature, trees can experience interruptions to their annual growth from such events as *drought*, unusual cold *weather*, or *defoliation* (e.g., by insects or hail); these interruptions of growth lead to many flaws and imperfections in the resulting wood.

Previously, we discussed the influence of grain on various strength and stiffness properties. While the wood grain is obviously not a defect (since it is the manifestation of the wood's fundamental nature), when a piece of lumber is cut at an angle to the wood grain, or the grain simply developed in an irregular manner, mechanical properties can be compromised. The misalignment of the grain and lumber cut may lead to reductions in tensile and compressive strength, corresponding bending strength, modulus of elasticity, and toughness.

Besides the grain, **knots** are probably the most commonly known physical property of wood and among the most potentially serious defects. A knot is the portion of a branch that has become embedded in the main trunk of the tree. The presence of a knot means that the grain of the trunk must either be interrupted or change direction to accommodate the intrusion of the branch. As a result, knots typically lead to a *reduction in mechanical property values* (tensile strength especially) in that section of the tree; stress concentrations build up, and checking (**figure 7.11**) is common around knots during drying.

FIGURE 7.11   Common wood defects: (a) Knots; (b) checks; (c) shakes.

Checks and **shakes** are both manifested by cracks in the wood, but checks tend to be across growth rings, while shakes tend to be parallel. **Figure 7.11(b–c)** help us to understand the orientation of checks and shakes, which is explained by how they were formed. When the shrinkage of a growth ring is significantly restrained by a knot or adjacent (underlying) growth ring, a split or check in the growth ring occurs, thus compromising the integrity of that ring. A loss of bonding between growth rings can occur when the tree has been exposed to prevailing winds during its growth, causing adjacent rings to slip apart and create a gap or crack (shake) parallel to the growth ring. In both cases, the result is a loss of strength in the wood.

Pitch pockets are openings in the wood (**figure 7.12**), typically between and parallel to annual growth rings, where free resin from the tree has been accumulated. Note that resin is not the same as sap; resin is stored in the outer cells of the tree, and when those cells are damaged, the resin is released to fill the damaged zone. Sap also derives from the sapwood, but serves a nutrient function, whereas the true function of the resin is unclear.

As one might expect, pitch pockets are typically confined to coniferous trees that we associate with high levels of resin, including pines, spruces, and firs, among others.

A defect in milled lumber that results from wood production rather than natural causes is a **wane**. This occurs when the board's cross-section is reduced by cutting it near the cambium or bark. Thus, rather than having a squared-off edge, a board with a wane will exhibit the rounded geometry of the tree's outer perimeter (**figure 7.13**). This is often accompanied by actual bark

FIGURE 7.12   Pitch pocket.

FIGURE 7.13   Wane.

remnants or a discoloration associated with the bark having been peeled from this outer layer. The most serious result of a wane is a reduction in cross-section, leading to less load-carrying capacity.

VII. Wood-Based Composites and Wood Treatment

As a natural, growing, biological organism, wood has structural timber limitations, including the size of sawn timber being limited by tree size and natural defects, which can limit allowable stresses. As a result, **wood-based composites** have been developed to produce different shapes with better mechanical properties than natural timber. To make these composites, we actually use wood waste and scrap wood that would normally be discarded, thus maximizing use of all parts of timber.

Panel Products

To create a *veneered panel* product, such as plywood, a log is debarked and then peeled in thin layers (**figure 7.14a**). These layers are glued together with grains of adjacent layers in a perpendicular manner, called cross-lamination. The advantages to this are very large sheets of wood, minimizations of wood defects, and isotropic properties for loading. Plywood is used in floors, walls, furniture, and for architectural purposes. To create *non-veneered panel* products, such as oriented strand board, large wood pieces are mixed with glue and pressed into panels with the grains clearly visible (**figure 7.14b**). Other products, such as particle board, are made from ground-up wood mixed with resin that is pressed into boards (**figure 7.14c**).

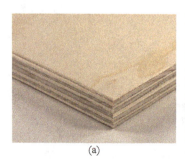
(a)

(b)

(c)

FIGURE 7.14   Wood panel products: (a) Veneered (plywood); (b) non-veneered (oriented strand board); (c) particle board.

Non-veneered panels are mainly used as subflooring or in furniture because of the lack of structural fibers available after their creation.

Glulam and Parallam

Glulam is another type of manufactured wood product that is created by gluing lengths of sawn lumber together with all grains running parallel to the long direction (**figure 7.15a**). The advantage to this construction is larger boards than can be milled from readily available trees (up to 130 feet long, 6 feet deep!), making glulam less expensive than sourcing large lumber. In many cases, stronger wood is placed in the outer positions with lesser quality wood in the middle, which places the best wood where the highest stresses are in bending.

Parallam is made from small strips of wood with cuts that are parallel to the grain, bundled, mixed with glue, and then pressed into wood pieces of desired dimension and shape (**figure 7.15b**). This method of wood composite creation reduces the effect of defects and ensures a perfectly parallel wood grain.

Manufactured Components

Increasingly in structural building, a number of specialized, composite-type wood materials have been used as structural components—replacing steel! This is done mostly through the use of proprietary (patented or single manufacturer) items, such as wood I-joists and other types of manufactured shapes (**figure 7.15c**).

Wood Treatment and Durability

For wood, its **durability** can be defined as the resistance of wood to attack from moisture, chemicals, organics (e.g., insects, fungi), and fire. Many of these durability problems with

(a)   (b)   (c)

FIGURE 7.15   (a) Glulam; (b) parallam; (c) manufactured components—note the groove for interlocking pieces.

wood can be resolved by kiln drying combined with a chemical treatment. **Pressure treatment** is when chemicals, such as water-soluble salts, are injected under pressure into wood pores. This type of wood is commonly used in moist locations away from actual human contact, as it is toxic to touch. The use of **creosote** is another method of wood preservation that employs coal tar-type compounds; it is commonly used on railroad ties. Lastly, paints, stains, and waterproofing chemicals (like Thompson's WaterSeal) are used to seal moisture out of wood. Sealants are the least reliable but safest for direct human contact.

VIII. Sustainability of Wood

Unlike other materials that we harvest from the earth (e.g., ores for manufacturing metals, fossil fuels), wood is naturally renewable based on the ability of new trees to grow in place of those that are harvested. However, using wood instead of metals or concrete does not mean our project is free from carbon emissions and other environmental impacts, or is inherently "sustainable." Like other materials, wood results in environmental impacts throughout its production, from harvesting to processing into final wood products (see **chapter 8**).

The durability of wood is also discussed in **chapter 8** and is one reason that wood is a valuable building material. In fact, after its useful life in a structure, wood has the advantage of being capable of reuse via physical breakdown (i.e., breaking the wood down in smaller pieces for other wood products and pulp for paper), or as an energy source (i.e., fuel for electric or heat generation). Unlike many other common materials, even if wood is disposed, it is non-toxic and naturally biodegradable given sufficient time.

Ultimately, with sustainable forestry and disposal practices in place, wood can be an excellent material with regard to sustainability.

IX. Concluding Remarks

In this chapter we discussed how wood is grown and cut, its imperfections, and how it behaves in loading. As one of the more traditional building materials, wood is plentiful and arguably one of the most sustainable. As technology improves, we are finding new uses for all wood harvested and are creating new engineering materials from an old source.

Please click on the link or use your cell phone to scan the QR code in order to see an example of durable wood, known as cross-laminated timber:

Wood Is the New Concrete?!?: Cross-Laminated Timber

https://popsci.com/article/technology/worlds-most-advanced-building-material-wood-0/

X. Problems

1. What is the difference between hardwood and softwood?

2. Explain how heartwood is different from sapwood. Understanding this difference, how deep would you need to drill a hole in an 18-inch diameter maple tree to get sap for making syrup?

3. A piece of oven-dried wood weighs 3.7 pounds and has a total volume of 113.8 inches^3. The specific gravity of the wood solids only is 1.5 (if you could compact the dry wood into a solid mass). The volumetric shrinkage of the wood, from green to oven-dried state, was 14.5 percent.
    a. What is the oven-dried specific gravity that should be reported?
    b. What should be the reported unit weight of the wood, in pounds/feet^3?
    c. What is the maximum moisture content of the green wood (if all wood voids were filled with water)?

4. A kiln-dried piece of wood has a moisture content of 13.2 percent and weighs 4.4 pounds. The kiln-dried specific gravity is 0.55.
    a. Determine the weight of wood solids.
    b. Determine the volume of the kiln-dried wood.
    c. If the volumetric shrinkage that occurs from the green state to the kiln-dried state is 5.4 percent, what was the volume of the green wood?
    d. What should be the reported unit weight of the wood?

5. A piece of Douglas fir has a moisture content just after cutting of 27 percent. After kiln-drying, the moisture content goes to 13.5 percent and its specific gravity is 0.55.
    a. For a 1-foot^3 piece of the kiln-dry wood, find the weight of the oven dry wood and determine how much it weighed when green.
    b. If the volumetric shrinkage during kiln drying was 11.4 percent, what is the reported unit weight of the wood?

6. Pressure-treated wood has preservative chemicals injected into the wood voids. Prior to pressure treating, a cut of wood in the kiln-dried condition had dimensions of 5.5 inches × 9.75 inches × 21 inches, it weighed 18.5 pounds, and the moisture content was 11.5 percent. The specific gravity of wood solids alone was 1.5.
    a. Assuming all voids could be filled with either water or chemicals, what volume of void space was available in the kiln-dried piece for the pressure-injected chemicals?

b. After the chemical injection, the wood is expected to swell. Based on the kiln-dried dimensions, the swell is expected to be 0.6 percent in the tangential direction, 0.4 percent in the radial direction, and 0.1 percent in the longitudinal direction. What will be the new dimensions of the board after pressure treatment?

7. Wood is typically three times as strong in tension as in compression when loaded parallel to the grain. Why don't we use more wood as tension members in structures?

8. Match the following wood defects with their definition.

Check	a. Cross-section of a branch intersecting the wood.
Wane	b. Outer ring splits, causing a discontinuity in the grain.
Shake	c. Deficiency in board geometry due to sawing into bark or cambium.
Knot	d. Longitudinal separation of growth rings due to wind during growth.

9. The figure (next page) shows an experimental set-up for testing the bending capacity of a wood crossbeam. Note that the grain directions for the two support columns are different. The vertical columns have a cross-section of 3 inches × 3 inches and are both 5 inches long. The beam has a cross-section of 2 inches × 4 inches and has a length of 12 inches. The stronger column has a compressive strength of 4,675 psi and a modulus of elasticity of 0.8×10^6 psi. The weaker column has a compressive strength of 1,545 psi and a modulus of 0.4×10^6 psi.

a. What is the maximum allowable load, P, on the beam if its modulus of rupture (MOR) is 5230 psi?

b. If a total load P = 28,000 pounds is supported, will either column fail due to this load? Show calculations to support your answer.

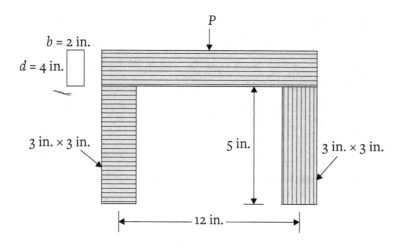

c. If only one-half the load of part (b) is put on the beam, what deformation will occur in each column under that loading?

10. Explain why the strength of most wood types is greater when loaded in parallel to the grain than when loaded perpendicular to the grain.

11. You are analyzing the bending strength of a wood I-beam. It is 1-foot high and is perfectly symmetric about its centroid, which is at the middle of the web. The area moment of inertia for this beam is 140 inches⁴. The tensile strength of the wood at the ends (the flanges) is 12,500 psi, and the compressive strength is 5,300 psi.

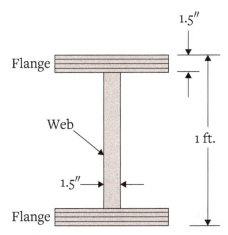

   a. What moment, in feet-pounds, will cause failure of this piece in bending based on the flange strengths?

   b. Will this fail in compression or tension?

12. You are assigned to test a new, multi-layered piece of wood. The wood consists of three layers of wood that are glued together. The two outer layers are made of wood A while the inner layer is made of wood B, with material properties listed in the table. You place an 8-inch-high, 6-inch-long (into the paper) piece of the material in a loading device with end plates <u>so that the same deformation is applied to all pieces</u> (figure right). Determine the individual strains at yield for the outer layers and the inner layer and the maximum load that can be applied without any of the wood failing.

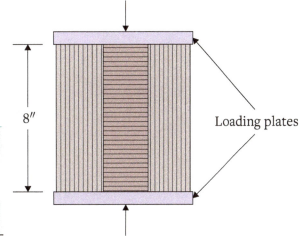

Loading plates

	Wood A	Wood B
Thickness (inches)	1.5	1.6
Compressive strength (psi)	4675	1545
Modulus of elasticity (psi)	0.8×10^6	0.4×10^6 psi

Credits

Fig. 7.1: Copyright © Coventry Log Homes (CC BY-SA 3.0) at https://en.wikipedia.org/wiki/File:Wood_growth_ill.jpg.

Fig. 7.3: Source: https://fog.ccsf.edu/~wkaufmyn/ENGN45/Course%20Handouts/14_CompositeMaterials/04_Wood.html.

Fig. 7.3a: Adapted from https://www.popularwoodworking.com/tricks/how-to-calculate-wood-shrinkage-and-expansion.

Fig. 7.10b: Copyright © Arturo Reina (CC BY-SA 3.0) at https://commons.wikimedia.org/wiki/File:Eixos.jpg.

Fig. 7.11a: Copyright © F pkalac (CC BY-SA 3.0) at https://commons.wikimedia.org/wiki/File:Wood_Knot.JPG.

Fig. 7.11b: Copyright © Beentree (CC BY-SA 3.0) at https://commons.wikimedia.org/wiki/File:Blue_stain_on_pinus_sylvestris_1_beentree.jpg.

Fig. 7.11c: Copyright © Minnecologies (CC BY-SA 3.0) at https://commons.wikimedia.org/wiki/File:Young_red_oak_with_decay.JPG.

Fig. 7.12: Copyright © Beentree (CC BY-SA 3.0) at https://commons.wikimedia.org/wiki/File:Resin_pocket_pinus_sylvestris_beentree.jpg.

Fig. 7.13: Copyright © by Virginia Wood Flows.

Fig. 7.14a: Copyright © Bystander (CC BY-SA 3.0) at https://commons.wikimedia.org/wiki/File:Spruce_plywood.JPG.

Fig. 7.14b: Source: https://commons.wikimedia.org/wiki/File:Chipboard_texture.jpg.

Fig. 7.14c: Copyright © Rotor DB (CC BY-SA 3.0) at https://commons.wikimedia.org/wiki/File:Particleboard.jpg.

Fig.7.15c: Copyright © VarunRajendran (CC BY-SA 3.0) at https://commons.wikimedia.org/wiki/File:Wood_plastic_composite.JPG.

8

Sustainability of Materials

I. Introduction to Sustainability

The materials that engineers employ must perform according to specifications, but ideally, we would also like them to be "sustainable." Sustainability is common to hear about nowadays, but what does it mean exactly? Unlike a strictly physical concept like hardness, sustainability is a broad term—there is no one single definition that applies to every situation; rather, there are many meanings, each appropriate to a particular context. One of the first and still most widely cited definitions appeared in a 1987 report titled "Our Common Future" by a UN Commission chaired by the then-prime minister of Norway Gro Brundtland. The report defined "**sustainable development** as development that meets the needs of the present without compromising the ability of future generations to meet their own needs."[1] Following from this definition, development might be labeled "unsustainable" if it decreases quality of life for future generations, for example, by using up scarce resources, permanently degrading landscapes and ecosystems, or releasing pollution that is toxic to humans and other organisms.

Sustainability is a vital concern for civil engineering materials, in part because of the sheer quantity of materials employed in constructing buildings and infrastructure. From 1960 up through the global financial crisis of the late 2000s, the production of basic materials went through an unprecedented expansion as industrialization spread and more people moved to cities, requiring new roads, bridges, buildings, and all manner of infrastructure. **Figure 8.1** shows this remarkable period of growth for the five major commodity materials. The increase in production in China, which now produces about half of the world's steel, was especially astounding: In the single year 2012–2013, the country *added* an additional 91 million tons of crude steel production

---

1 World Commission on Environment and Development (1987) "Our Common Future" Oxford: Oxford University Press.

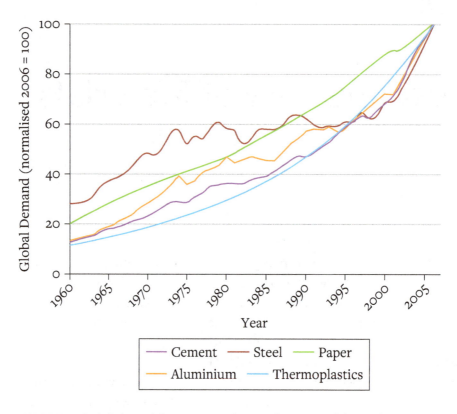

FIGURE 8.1 Period of growth for cement, steel, paper, aluminum, and thermoplastics.[2]

capacity, more than the entire existing output of the United States, the world's third-largest producer! Producing materials obviously requires using natural resources and leads to various types of emissions, some benign, but some not. Despite the fact that production technology has improved over time, leading to higher efficiencies and less resource use and emissions *per unit*, the tremendous scale of global industry ensures that the effects are widespread and can be damaging to surrounding communities and environments—in other words, unsustainable.

So, what does a sustainable material look like? A commonly used set of sustainability criteria is called the **triple-bottom line**. First, and most obviously, materials must be affordable to consumers and turn a profit for their producers. Silver is the most conductive element, but it would be ludicrously expensive to wire a home with silver in order to reduce resistance losses and save electricity; instead, we use copper, the second-best conductor. This is the conventional **economic** bottom line. Second, materials must be produced in

2   Julian M. Allwood, Michael F. Ashby, Timothy G. Gutowski and Ernst Worrell, "Material efficiency: A White Paper," *Resources, Conservation, and Recycling* 55, no. 3 (2011): 362–381.

a manner that benefits local communities and respects fair labor practices and human rights. For example, workers in factories should be of legal working age and should have access to personal protective equipment if they work in conditions that are hazardous to their health. These are criteria of social equity or the **social** bottom line. Finally, materials should be produced in a way that does not permanently deplete natural resources or degrade environmental quality. Ideally, we would prefer if our material economy could be based on renewable or earth-abundant resources, processed using renewable energy, consumed no net water, and released zero harmful substances to the environment. Even better would be materials that have a net positive effect on the environment. In reality, we are a very long way from this, but these types of considerations fall under the umbrella of the **environmental** bottom line. Taken together, these are the three components of the triple-bottom line, or the "three pillars of sustainability."

As with a three-legged stool, if you take one leg away, the result is collapse. A strategy of protecting human health and the environment without any consideration of cost will drive companies out of business, leading to reduced economic opportunities and affluence, and even eliminating our capacity to produce the technologies that underpin our modern way of life. This scenario is "bearable" but does not allow people the financial capacity to provide well for themselves and their families. **Figure 8.2** shows the areas of overlap among considerations of people, planet, and profit. Only where all three are included can a material, product, or project be considered sustainable for all stakeholders.

This is a high-level discussion, meant to introduce you to some of the theoretical foundations of sustainability. In practice, evaluating the sustainability of materials and using the results to guide engineering design and decision making is much more complex than the simple Venn diagram in **figure 8.2** and is often fraught with direct conflicts among sustainability goals. For civil engineers working in public agencies or private firms, many social and economic sustainability goals are achieved through typical practices, such as complying with codes and regulations, and avoiding cost overruns. Environmental sustainability, on the other hand, is not at all assured, although certifications and standards for materials or even entire infrastructure projects are developing

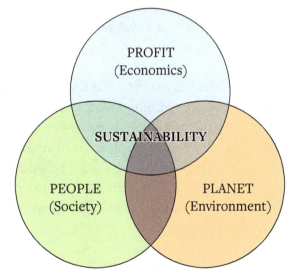

FIGURE 8.2   Venn diagram of sustainability.

rapidly. In the rest of this chapter, you will learn about some of the most common measures of environmental sustainability and their application to civil engineering materials.

II. Energy Flows and Embodied Energy

Energy is a fundamental indicator of environmental sustainability. The discovery of new energy sources and methods of harnessing that energy has been a critical thread in the development of human civilization. Our modern way of life would be impossible without inexpensive and accessible energy. Our energy infrastructure literally sustains our economies and our societies, but the extraction and use of energy has had very serious consequences for the natural environment and, by extension, for our health and well-being.

As the global population has grown and industrialization has spread, energy use has risen dramatically. In the half century from 1965–2015, the population more than doubled, from 3.3 to 7.3 billion people, but primary energy use nearly quadrupled to 13.1 billion tons of oil equivalent in 2015, mostly supplied by fossil fuels.[3] Extracting all of this fossil energy has come at a heavy environmental cost through degradation of terrestrial and aquatic systems via the removal of entire forests, oil spills choking marine ecosystems, and mountains decapitated for the coal they contain. Burning fossil fuels is also the dominant contributor to air pollution, causing the disruption of natural systems on a global scale and millions of premature deaths annually.[4] Renewable forms of energy, such as wind, solar, and hydropower, do not contribute directly to air pollution but have environmental consequences of their own, such as damning of rivers and large-scale alteration of landscapes. Because of its central position in our technological age, energy is a critical input to both economic and social development and is correlated with many other metrics of environmental impacts.

Laws of Thermodynamics and Energy Conversions

Our use of energy and, more broadly, the conversion of energy from one form to another is governed by the laws of thermodynamics, a branch of physics. The first law of thermodynamics states, "Energy can neither be created nor destroyed," which underpins the familiar concept of energy balances in a system used in fields across science and engineering. For example, burning coal in a cement kiln converts energy contained in the chemical bonds of the coal to an identical quantity of energy in the form of heat delivered to the kiln.

---

3   "Statistical Review of World Energy," British Petroleum, 2019, https://www.bp.com/en/global/corporate/energy-economics/statistical-review-of-world-energy.html.

4   "Global, Regional, and National Age-Sex-Specific Mortality for 282 Causes of death in 195 Countries and Territories, 1980–2017: A Systematic Analysis for the Global Burden of Disease Study 2017," *The Lancet,* 392, no. 10159 (2018): 1736–1788.

The second law of thermodynamics states, "The total entropy of an isolated system increases over time." Entropy is essentially a measure of disorder of a system. So, another way of expressing the law is that "the overall order of a system decreases over time." The law applies to any physical system but is especially useful when thinking about energy. Coal is a highly ordered collection of atoms. Electricity is a highly ordered collection of electrons. When these highly ordered energy carriers undergo a conversion to another form of energy, the second law mandates that the energy that comes out is less ordered than the energy that goes in, or of a lower *quality*. The first law mandates, on the other hand, that the total *quantity* of energy remains the same in the system. Lower quality energy means that less of that energy can be harnessed for doing work.

In practical terms, what this means is that all energy conversions entail losses of useful energy. Losses frequently take the form of friction or heat. For example, a light bulb converts electrical energy to radiation in the form of visible light but also produces some heat. At a larger scale, a car with an internal combustion engine converts chemical energy contained in the gasoline into kinetic energy but with losses to wind friction, road friction, heat contained in the exhaust, even noise (a form of mechanical energy).

Now let's consider an entire economy. Since the 1970s, Lawrence Livermore National Laboratory (part of the US Department of Energy) has created diagrams that show an annual energy balance for the entire country. **Figure 8.3** shows the energy balance for 2015 in units of "quads," or quadrillion British Thermal Units (BTU). Following the figure, from left to right, we see types of energy supplies, conversion into electricity, and the four major energy end-use sectors of residential, commercial, industrial, and transportation, all connected by energy flows. Energy supplied in raw form is called **primary energy**, such as solar radiation or coal extracted from a mine. Primary energy may then be cleaned, transported, and/or converted into other intermediate forms of energy, called **secondary energy**. The most important secondary energy conversion is from primary energy into electricity. A quick look at **figure 8.3** shows that in 2015 the United States used 38 quads of primary energy to produce 12.6 quads of electricity, meaning that 25.4 quads (or 67 percent) was lost to heat (or "rejected energy" in the figure). Energy delivered to the four end-use sectors is called **final energy**. Each end-use sector uses final energy to deliver what we actually want from energy, which is **energy services**. This is the conditioned air, formed steel, or transportation of materials from point A to point B that we use in our homes and businesses. Note that the energy conversions taking place in each end-use sector also produce losses and that the losses can be quite large—nearly 80 percent in the case of transportation. Taken as a whole, only 39 percent of the primary energy that is mobilized by the US economy actually does anything useful for us. The majority (61 percent) is practically lost to friction and heat.

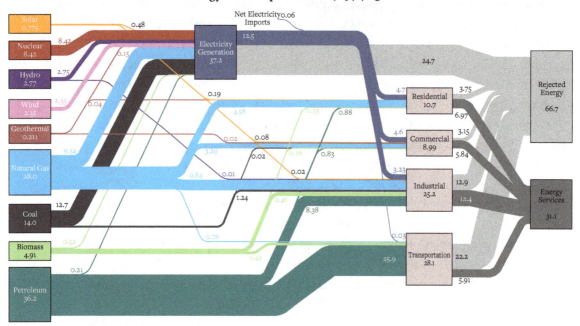

FIGURE 8.3 Energy flows in the US economy in 2017. Units are "quads" of energy, or quadrillion BTUs.

Embodied Energy of Materials

The production of civil engineering materials takes place in the industrial end-use sector depicted in **figure 8.3**. The industrial sector includes the mining and processing of minerals and ores, refining into materials, forming into products, and delivering to the job site. As civil engineers, we may know the last step of a material's journey, the distance that steel travels from the fabricator or concrete from the batch plant to the job site. We could use this distance to calculate how much energy it took to transport that material, but a broader measure of sustainability would be the energy it took to produce the material from start to finish, from excavation all the way to installation in buildings and infrastructure. This is called the **embodied energy** of a material.

Embodied energy reflects several kinds of energy use. Using steel as an example, mining and crushing iron ore requires diesel fuel; smelting typically uses natural gas, coal, or coke for heat; alloying may be done in an electric arc furnace (EAF); and milling machines run on electricity. When adding up energy use across all these processes, it is critical that we add the same types of energy together. We cannot simply add 1 MJ of natural gas use with 1 MJ of electricity use, because producing and delivering that electricity might itself require 3 MJ of natural gas because of the conversion losses discussed earlier. For this reason, embodied energy is always expressed in primary energy terms, and other types of energy

EXAMPLE 8.1: PRIMARY ENERGY CALCULATIONS

You are in an area where electricity is predominantly produced from natural gas. Extracting and transporting natural gas has losses of 5 percent, the gas-fired power plant has an efficiency of 45 percent, and the losses of in the electrical infrastructure are 2 percent for AC/DC conversion, 7 percent in transmission and distribution, and 3 percent in wiring. Suppose you upgrade your refrigerator from an 18–cubic foot unit that uses 500 kWh per year to a larger 25–cubic foot unit that uses 350 kWh per year.

How much <u>final</u> energy are you saving? How much primary energy?

SOLUTION

Final energy is the electricity delivered to "end uses," in this case, your refrigerator. This is simply the difference in annual electricity use, or 500 – 350 = 150 kWh.

Primary energy is expressed in terms of the raw energy source, which is natural gas in this region. The common symbol for energy conversion efficiency is η, the Greek letter "eta." First, we must express each energy conversion step in terms of its efficiency:

- Natural gas well $\rightarrow$ natural gas at power plant η_1 = 95 percent
- Natural gas at power plant $\rightarrow$ DC electricity at power plant η_2 = 40 percent
- DC electricity at power plant $\rightarrow$ AC electricity at transmission start η_3 = 98 percent
- Electricity transmission start $\rightarrow$ Electricity at household meter η_4 = 93 percent
- Electricity at household meter $\rightarrow$ Electricity at refrigerator η_5 = 97 percent

Primary energy can be calculated from final energy based on all of the intermediate conversion efficiencies as

$$E_{PRIMARY} = \frac{E_{FINAL}}{\prod_i \eta_i}.$$

So, the primary energy saved by upgrading your refrigerator is

$$E_{PRIMARY} = \frac{150 \text{ kWh}}{\eta_1 \cdot \eta_2 \cdot \eta_3 \cdot \eta_4 \cdot \eta_5} = \frac{150 \text{ kWh}}{0.95 \cdot 0.40 \cdot 0.98 \cdot 0.93 \cdot 0.97} = \frac{150 \text{ kWh}}{0.346} = 447 \text{ kWh}.$$

Actually, it is even better than this, because you are getting a larger refrigerator! If we consider energy use in terms of services being provided, the older fridge has a normalized final energy use of 27.8 kWh/cubic foot, while the newer fridge has a value of 14 kWh/cubic foot, which translates to an annual primary energy savings of

$$E_{PRIMARY} = \frac{13.8 \text{ kWh / cu.ft.}}{0.346} = 41.1 \text{ kWh / cu.ft.}$$

further down the chain (secondary or final energy) must be translated back into primary energy.

Embodied energy has been calculated for hundreds of materials, including the major civil engineering materials listed in **table 8.1**. It is important to note that, because conversion losses depend on the mix of energy technologies being used in a particular region (such as coal-fired power plants versus wind farms for electricity), embodied energy values depend on where a material is produced. That is, embodied energy is a **circumstantial** sustainability metric (depending on the circumstances of energy and material production) rather than an **inherent** sustainability metric, such as toxicity that depends only on a material's physical or chemical properties.

Table 8.1 lists embodied energy per unit mass of material (MJ/kg). When comparing materials, however, it is important that we compare based on their function—that is, how much material would actually be used for a specific application. For example, carbon fiber composites have a higher tensile strength than all but the strongest steels, and so less actual material is needed for tensile applications. This is the same consideration we would have in a comparative cost analysis: We have to compare the total material bill rather than the per unit costs. **Example 8.2** shows how embodied energy values can be scaled to an appropriate **functional unit**, or desired functional specification. Note that this is analogous to normalizing by the volume of the refrigerator in **example 8.1**.

The world of materials is vast, and material scientists are constantly adding new options for design engineers to choose from. Ideally, designers would be able to consider simultaneously both performance characteristics of materials and environmental metrics, such as embodied energy. British engineer Michael Ashby has developed a visualization tool to enable just such a juxtaposition. So-called Ashby plots, an example of which is shown in **figure 8.4**, are rich in information and allow designers to differentiate both between and within families of materials and to choose the least energy-intensive option. For example, **figure 8.4** shows that both carbon steels and nickel alloys have similar yield strengths, but carbon steels are nearly an order of magnitude more energy intensive to produce on a volumetric basis.

III. Emissions from Material Production

Embodied energy is an important measure of sustainability, especially when considering much of the energy used to produce materials is sourced from nonrenewable sources, such as petroleum. But oftentimes, we are more concerned with the emissions that result from burning fuels for heat and electricity than with the actual use of the fuels per se. Just as we

TABLE 8.1 EMBODIED ENERGY OF MAJOR CIVIL ENGINEERING MATERIALS

Material	Embodied Energy MJ per kg	Embodied Carbon kg CO_2-e per kg
Aggregate	0.083	0.005
Aluminum (virgin)	218	8.70
Aluminum (recycled)	29.0	1.75
Asphalt (5 percent binder)	3.39	0.067
Bricks (common)	3.00	0.24
Cement (Portland, 94 percent clinker)	5.50	0.94
Cement (35 percent fly ash)	3.68	0.62
Cement (35 percent blast furnace slag)	4.21	0.65
Concrete (Portland cement, 16–20 MPa)	0.70	0.10
Concrete (Portland cement, 40–50 MPa)	1.00	0.15
Concrete block (medium density)	0.67	0.08
Concrete, steel reinforced (25–30 MPa, 110 kg rebar per m³ concrete)	1.92	0.20
Copper pipe (virgin)	57.0	3.73
Copper pipe (recycled)	16.5	0.82
Glass	15.0	0.89
Insulation, fiberglass	28.0	1.35
Insulation, polystyrene (XPS)	88.6	2.92
Insulation, polyurethane (rigid)	101.5	3.87
Iron	25.0	1.97
Sand	0.081	0.005
Soil, compressed	0.45	0.02
Steel, carbon (virgin)	35.4	2.80
Steel, carbon (recycled)	9.40	0.46
Steel, stainless	56.7	6.15
Stone	1.26	0.08
Timber, hardwood (excluding sequestration)	10.4	0.24
Timber, softwood (excluding sequestration)	7.4	0.20
Timber, plywood (excluding sequestration)	15.0	0.44
Timber, glue laminated (excluding sequestration)	12.0	0.41

Note: Inventory of carbon and energy v.2.0; embodied carbon values are midpoints of reported ranges.

EXAMPLE 8.2: COMPARING STEEL VERSUS CONCRETE

Let's make the classic comparison between a steel versus reinforced concrete structure. Suppose that the primary function of our simple structure is to withstand compression. The compressive strength of low-alloy steel is ~150 MPa, while the value for high-strength Portland cement concrete is ~50 MPa. Which type of structure will have the lower embodied energy?

SOLUTION

Based on these strength values, for structures only under compression, we will need three times as much concrete as steel on a volumetric basis. The embodied energy values in **table 8.1** are listed as 35.4 MJ/kg for virgin low-alloy carbon steel and 1 MJ/kg for concrete. These values are given per unit mass, so we must convert using the density of these materials:

$$\rho_{\text{STEEL}} = 8{,}000 \frac{\text{kg}}{\text{m}^3},$$

$$\rho_{\text{CONCRETE}} = 2{,}400 \frac{\text{kg}}{\text{m}^3}.$$

So, the ratio of embodied energies of two *functionally equivalent* structures will be

$$\frac{\text{En Embodied}_{\text{STEEL}}}{\text{En Embodied}_{\text{CONCRETE}}} = \frac{1}{3} \cdot \frac{8{,}000 \frac{\text{kg}}{\text{m}^3} \cdot 35.4 \frac{\text{MJ}}{\text{kg}}}{2{,}400 \frac{\text{kg}}{\text{m}^3} \cdot 1 \frac{\text{MJ}}{\text{kg}}} = \frac{283{,}200 \text{ MJ}}{7{,}200 \text{ MJ}}.$$

Even though steel has a much higher compressive strength than concrete, because of its higher density and embodied energy per kilogram, the steel structure is much more energy intensive to produce. Note that this still holds even if we use 100 percent recycled steel.

---

add up all energy used to produce a material to find its embodied energy, we can also add up all of the emissions that occur to produce that same material. In general, this is called the **embodied emissions** of a material. However, usually, we want to be specific to a single type of emission. The most well-known example of this is embodied greenhouse gas (GHG) emissions, for gases such as carbon dioxide and methane that contribute to the greenhouse effect and global climate change. Another name for this measure is the **carbon footprint** or **embodied carbon** of a material. Values for embodied carbon are listed alongside embodied energy values for major civil engineering materials in **table 8.1**.

Like embodied energy, carbon footprint considers emissions over the life cycle of a material, starting with extraction of raw materials and including processing and refining,

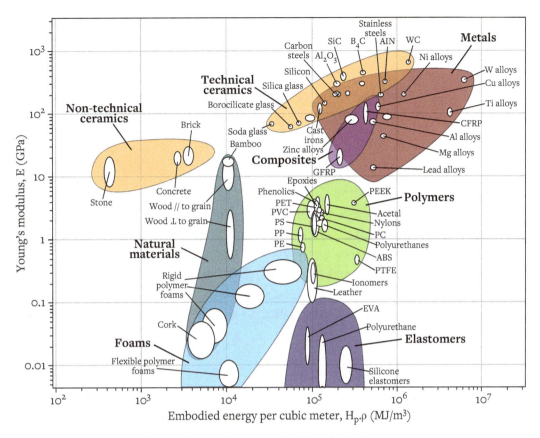

FIGURE 8.4 "Ashby" plot of material strength versus embodied energy, log-log.

forming into products, and transportation. Globally, the majority of GHG emissions occur from combusting fossil fuels, but GHG emissions can also arise directly from industrial or agricultural processes. Three important sources of GHG emissions overlap with the major civil engineering materials considered in this book: steel furnaces, cement kilns, and forestry operations.

Steelmaking

In basic steelmaking, the first step, carried out by a blast furnace, is to chemically reduce the iron oxide (Fe_2O_3) to molten iron (Fe). This is accomplished when oxygen-laden blast air reacts with metallurgical coke (solid carbon) to produce carbon monoxide, which in turn reacts with the iron oxide to reduce it to pure iron:

$$2C + O_2 \rightarrow 2CO,$$

$$Fe_2O_3 + 3CO \rightarrow 2Fe + 3CO_2.$$

Some of the resulting carbon dioxide is re-reacted with coke to produce carbon monoxide, but at the end of the process, all of the CO_2 is eventually released to the atmosphere. In addition, limestone (calcium carbonate, $CaCO_3$) is added to blast furnaces in order to remove impurities in the iron, such as silica. This reaction also produces CO_2 emissions as the limestone is thermally decomposed to produce lime (calcium oxide, CaO):

$$CaCO_3 \rightarrow CaO + CO_2,$$

$$SiO_2 + CaO \rightarrow CaSiO_3.$$

Adding these "process emissions" to the emissions from fuel combustion that provides heat to the process gives an average emissions factor of 1.8 tons of CO_2 emitted per ton of steel produced.[5] You'll see this is lower than the embodied carbon value for "steel, carbon (virgin)" in table 8.1. This is because the embodied carbon value also includes emissions that happen outside of the steelmaking process, including mining and processing iron ore, quarrying limestone, and transporting these minerals to the steelmaking site. In addition, embodied carbon counts not just carbon dioxide emissions, but also emissions of other GHGs such as methane.

Cement Kilns

During Portland cement production (the most common type of cement), limestone is heated with materials, such as clay, to produce calcium silicates. As in the blast furnace, thermal decomposition of the limestone produces carbon dioxide:

$$CaCO_3 \rightarrow CaO + CO_2.$$

The production of 1 ton of cement gives rise to just under a ton of CO_2 emissions, which are nearly evenly split between process emissions and emissions from fuel combustion.

Due to the sheer scale at which we produce steel and cement, they are very important sectors in terms of overall global GHG emissions. The iron and steel sector contributes nearly 7 percent of global CO_2 emissions, while the cement sector contributes approximately 5 percent.[6]

In addition to carbon dioxide emissions, the production of steel and cement also give rise to other types of pollutants to air, water, and soils. For example, sulfur that is present as a contaminant in iron ore and cement-making minerals, or in the fuels used to provide heat, may be oxidized to sulfur dioxide species (commonly known as SO_2), which contributes to acid rain. High-temperature processes that occur in the presence of air will produce nitrogen

---

5   World Steel Association, http://www.worldsteel.org/steel-by-topic/life-cycle-assessment.html.

6   International Energy Association (IEA), https://www.iea.org/.

oxide species (or NO$_x$) that are dangerous to human health. Metal contaminants, such as arsenic, mercury, nickel, and vanadium, may also be released in the air from industrial processes in the form of small particles (or particulate matter) that are a significant cause of human disease and mortality worldwide.

Forestry Operations

Wood products are a completely different story. As you learned in **chapter 7,** wood is sometimes touted as a more sustainable building material than steel or concrete. The main reasons for this are (1) that wood is a renewable resource that can be replanted and harvested repeatedly and (2) that wood sequesters carbon dioxide from the atmosphere through photosynthesis, thus helping to mitigate climate change. While these are both true, it is *not* the case that wood is emissions free! As usual, the story is more complicated. To produce lumber, wood must be harvested, transported, debarked, and/or milled to size. Each of these steps requires energy and produces emissions from fuel combustion. Some forestry operations apply nitrogen fertilizers to promote tree growth. These fertilizers themselves require energy to produce and can be converted by natural processes to nitrous oxide (N$_2$O), which is a potent GHG. New roads must typically be created in order for the harvesting equipment to gain access, which can lead to erosion and clog local waterways with sediment. Treating wood through surface treatments or pressure treating or manufacturing wood composites, such as glulam and plywood, relies on chemicals, some of which can be hazardous to environmental or human health and which can be emitted during production or released during use.

The sequestration of carbon dioxide is not the only way in which forestry operations affect the carbon balance. A significant portion of the carbon sequestered by forests is below ground, in soils and root structures. After trees are harvested, if the stumps are pulled out and chipped or burned, and the areas are tilled prior to replanting, then most of the soil carbon will escape, including carbon that was sequestered by organisms other than the trees. In order to promote sustainability in the wood products industry, many companies and owners are managing their forestry operations in order to minimize emissions, erosion, and impacts on wildlife, while maximizing the carbon that stays in soils. Several certification systems and organizations exist, such as the Forest Stewardship Council and Sustainable Forestry Initiative that ensure forests are being managed and allow wood from those forests to be marketed with an eco-label, which tends to command higher prices.

There is more to trees than lumber. The branches and crowns (or "lops and tops") of trees can also be used to make value-added products. If these coproducts substitute for energy or materials that would otherwise require the use of virgin resources, then, again, we can reduce the amount of energy use and emissions overall, which provides further environmental benefits to wood as compared to other engineering materials.

IV. Making Materials More Sustainably

Industrial producers of materials are well aware of the environmental implications of their operations, and many have made impressive strides in reducing energy use, improving material efficiency, and using more recycled materials. Although making these improvements often requires significant financial investments, companies can save money in the long term through reduced energy and material costs. Here we summarize a few of the most promising technologies and trends.

Using Heat Wisely

Much of the energy used in steelmaking and cement production is for heating, so one of the most fruitful strategies for reducing energy use in these industries has been to take waste heat from one part of the process and use it in another. This general strategy is called **energy integration**. For example, hot blast furnace gases in a steelmaking complex can be run through a waste heat boiler to generate high-pressure steam, which can then be used in other mill operations.[7]

Another strategy has been to consolidate the heating and cooling process in order to avoid heating materials up that have just been cooled. Traditionally, steel was cast into individual ingots, allowed to cool, and then reheated and shaped into semi-finished products. The advent of continuous casting in the 1950s allowed for steel to be shaped directly into semi-finished products. These would then be rolled into final products through hot or cold rolling, which would again require reheating. More recently, integrating casting and rolling (or "direct casting") connects the casting and hot rolling operations so that reheating is avoided, and steel can be rolled directly into final products as it is cooling.

Alternate Energy Sources

Natural gas, coal, and coke are the most common fuels used to supply heat for steel furnaces and cement kilns, as well as a carbon source for the chemical reactions that occur (see section III). In many facilities, waste materials and alternate energy sources are being used that substitute for a portion of these nonrenewable fossil fuels. Examples include waste tires, oils, sludges, and even postconsumer waste plastics. Another important trend for switching energy sources has been the adoption of electric arc furnaces (EAF) in

---

7   Ernst Worell et al., "Energy Efficiency Improvement and Cost Savings Opportunities for the U.S. Iron and Steel Industries," *Energy Star*, https://www.energystar.gov/ia/business/industry/Iron_Steel_Guide.pdf?25eb-abc5).

steelmaking. In an EAF, metal is melted by passing electrical currents through it rather than by combusting fuels to heat the furnace and is more energy efficient than traditional blast furnaces. In addition, if the electricity is produced by renewable means, such as hydropower, then the use of EAFs can greatly reduce emissions associated with steelmaking. EAFs can also process high proportions of recycled steel scrap, which is more difficult to do in a traditional furnace.

Using a Higher Proportion of Recycled Materials

Substituting secondary materials for primary materials avoids the embodied energy and emissions from virgin resource extraction and processing. Increasing recycled content can also save on material costs and ensure a local supply of material. So why aren't recycled content percentages higher? One major reason is concerns about contamination, which can degrade product quality and lead to materials being out of specification.

Recycled materials, especially postconsumer, tend to be highly heterogeneous. For example, carbon steel scrap may also contain several other metals that act as coatings or alloying elements, as well as materials such as copper piping that weren't separated during scrap processing. In order to produce steel to specifications, remelters can conservatively control the composition by diluting scrap metal with virgin direct-reduced iron. With more sophisticated sensing and control technologies, remelters can now balance their charges more effectively and so can accept a higher proportion of scrap metal.

Recycled materials are also commonly used in concrete production. Ash has been used in concrete for its pozzolanic properties since Roman times. However, rather than using natural volcanic ash, now we make use of ash emitted from coal combustion. Fine particles emitted from burning coal in electric power plants are called **fly ash**. These particles are light enough to be carried with the flue (exhaust) gases out of the boiler but are largely captured by pollution control equipment. Fly ash particles are primarily aluminosilicates but can also contain calcium oxide and can substitute typically 30 percent of the Portland cement used in concrete. Processing the fly ash before use can allow for even higher substitution percentages. Other materials that can substitute for cement include steel blast furnace slag and fumed silica from the refining of silicon for electronics. Again, because of the high embodied energy and emissions from the production of cement, substituting with a recycled (and lower-impact) material can improve the environmental sustainability of the resulting concrete. Recycled aggregate can also be used in concrete and avoids having to dispose of this material. As of 2004, 11 states allow for direct recycling of old concrete into new Portland cement concrete.[8]

---

8 "Recycled Aggregates," Portland Cement Association, https://www.cement.org/learn/concrete-technology/concrete-design-production/recycled-aggregates.

V. Sustainability During Use: Material Longevity and Thermal Performance

A material's environmental impacts do not end once it is incorporated into buildings and infrastructure. Another important perspective of material sustainability is how it performs while in use. Of course, performance is what we are ultimately after when choosing a material, but there are certain performance considerations that also affect the environment. In this section, we will concentrate on two of the most important considerations for civil engineering materials—namely, material longevity and thermal performance.

Material Longevity

We want our materials to last, but for how long? Engineering a structural material to last indefinitely, under a wide range of conditions, is technically challenging and therefore costly. Materials undergo physical degradation because of cyclic or extreme loads, or chemical degradation from reactions that occur in natural environments, and this degradation will eventually have a deleterious effect on material performance. Based on the typical rate of degradation of a material in a given application, we can assign it a material longevity, or time to failure. Assessing all material components of a civil engineering project together allows engineers to assign a design life and plan when maintenance or replacement must be carried out. Different materials have different rates of physical and chemical degradation, and therefore different lifetimes. A material with a short lifetime may have to be replaced multiple times over the course of a project's life (for example, the surface pavement of a highway). This is especially important for civil engineering projects, which tend to have much longer design lifetimes than other types of engineered products and are frequently used far beyond these design lifetimes.

FIGURE 8.5   Visible corrosion of civil engineering materials.

For civil engineering materials that are typically exposed to outdoor environments, the most important degradation mechanism is corrosion. **Corrosion** is a chemical process by which gaseous components of the atmosphere (such as oxygen) or liquid (such as sulfuric acid in raindrops, or water itself) react with the exposed material surface. For metals such as steel, this is typically an oxidation reaction with oxygen or sulfur playing the role of oxidant. The most familiar example of this is the oxidation of iron to iron oxide, or common rust (**figure 8.5**). Other metals commonly used in structures, such as aluminum and copper, can also undergo corrosion to aluminum oxide and copper oxide, with recognizable results.

Concrete can also corrode, or, more accurately, the steel rebar reinforcement can undergo corrosion. This can occur through chloride ingress, when chloride on the surface of a concrete structure (from sea spray or road salt) will diffuse through to the rebar and react. Rebar can also corrode because of carbonation, when the calcium hydroxide in the cement essentially absorbs carbon dioxide from the atmosphere to reform calcium carbonate and release water. This changes the pH of the pore water in the concrete, making the rebar more susceptible to corrosion once the carbonation front reaches the reinforcement.

The economic costs of corrosion are staggering. Research by National Association of Corrosion Engineers (NACE) International puts the costs of corrosion of structures, infrastructure, and equipment at more than $450 billion in 2013 dollars, or 2.7 percent of the US gross domestic product. The global costs are estimated to top $2.5 trillion.[9] Much effort and expense, therefore, goes into protecting materials from corrosion. For example, a layer of zinc is commonly added to the surface of steel in a process called **galvanization**. Oxygen in the air or dissolved in water then reacts with the zinc instead of with the iron in the steel, which consumes the zinc (it is called "sacrificial") but prevents rust from occurring. Another strategy is to specify materials that are less vulnerable to corrosion. For example, stainless steel, containing chromium and frequently nickel, does not readily corrode and is often specified in high-value applications where corrosion would occur readily, such as chemical plants. **Passivation** is another strategy to protect against corrosion, where a thin layer of oxide or nitride is allowed to form on the surface of a metal, protecting it against further corrosion in the interior. Passivation layers work as long as the metal oxide or nitride cannot readily dissolve into the surrounding environment. Steel rebar in concrete forms a passivation layer because iron oxide dissolves extremely slowly in the alkaline environment of the surrounding concrete. (Carbonation lowers the pH, allowing metal ions to dissolve and causing this passivation layer to fail.)

Other environmental conditions can also affect the longevity of civil engineering materials. Changes in temperature and humidity affect the action of water within cementitious materials, particularly when temperature swings through freeze-thaw cycles and cause crack formation and propagation because of migration and expansion of ice. This is one of the main causes of failure in concrete pavements. Air pollutants, such as SO_2 and NO_x, react with water in raindrops to form sulfuric and nitric acids, or acid rain. These acids can dissolve minerals in stone and concrete structures, mortars in masonry structures, and metals, causing a loss of mass.

---

9   NACE International Impact Study, 2016. http://impact.nace.org/.

For wood, the main degradation mechanisms are physical weathering and biological attack. Few microbes or animals are able to make use of wood for food, even though it contains a relatively high energy potential through its sugar stores. Of course, this resistance to breakdown is the same reason that wood continues to be such a valued building material—it resists and is therefore not attractive to more organisms that might use it for food. The difficulty of chemically breaking wood down for reuse is known as biomass recalcitrance and is thought to be caused by a variety of factors, but primarily because of the complexity of cell-wall constituents, the degree of lignification or adhesion among cells, and resistance to chemical processes that might normally break down the cells.

What does all this have to do with sustainability? As you learned in the preceding sections, the production of materials requires energy (and water and other nonrenewable resources) and results in emissions to air, water, and soils. If a material must be replaced five times, this is five times the energy and five times the emissions that must be accounted for in our sustainability comparisons across material options (to say nothing of the additional economic costs). Protections against corrosion, such as galvanizing steel, require additional materials and energy that must also be considered.

Thermal Performance

In the summer and winter months, one prime function of buildings is to provide a comfortable temperature for occupants, through heating and cooling. The materials that have been used to construct the building play a large part in how efficient that building is in keeping heat in or out. Heating and cooling require energy, either in the form of heating fuels or electricity for running heat pumps or air conditioners, for example. As you learned in section III, saving one unit of electricity might save three to four times that amount of primary energy, depending on how the electricity was generated. Therefore, an important consideration in selecting a sustainable material is how it will influence the amount of energy that a building uses.

Heat in a warm interior in the winter naturally tends to move to the cold exterior, and vice versa in the summer months. How fast heat moves through a material is governed by its **thermal conductivity** k, with SI units of watts per meter per degree Kelvin. (This is analogous to the electrical conductivity σ, which indicates how easily electrons can move through a material.) Materials with a high thermal conductivity are able to transfer heat quickly from hot to cold areas. Anyone who has burned themselves on a frying pan handle could easily guess that metals such as steel and aluminum are good thermal conductors. Concrete and particularly wood, on the other hand, are relatively poor conductors, otherwise known as **thermal insulators**. **Table 8.2** shows thermal conductivity values for common civil engineering materials.

Heat loss through an engineering material depends both on thermal conductivity and on the cross-section of material needed. Imagine a simple house with walls, floor, and roof made out of a single material of thickness x, and a difference between internal and external temperatures of ΔT. The rate q at which heat is lost through an area A of wall is given by the law of heat conduction, known as **Fourier's law**:

TABLE 8.2 THERMAL CONDUCTIVITY OF SELECTED MATERIALS

Material	k (W/m-K)
Aluminum	205
Iron	80
Steel	50
Concrete	0.8
Glass	0.8
Masonry brick	0.6
Wood	0.04–0.12
Rock wool (insulation)	0.04
Air	0.03

$$q = \frac{kA\Delta T}{x}.$$

Now, walls are typically assemblies of different materials, all with different thermal conductivities. Luckily, it is straightforward to calculate the thermal conductivity of an assembly of materials based on the individual conductivities of its material components. These relations will look familiar, as they are analogous to the rules used to combine Hookean springs in series or in parallel. First, let's define a new term, U value, as

$$U = \frac{k}{x}.$$

This value simply normalizes the thermal conductivity of a material k by its thickness. Another common measure is R value, which is just the inverse of U value:

$$R = \frac{1}{U}.$$

(For those of you familiar with building materials, R and U values are commonly used to rate insulation or windows.) So, the flow of heat through a single material is just:

$$q = UA\Delta T = \frac{A\Delta T}{R}.$$

Now, suppose we have n materials stacked on top of each other and heat is being conducted through the stack, normal to the inside or outside faces (**figure 8.6**). We can treat this as an example of Hookean springs in series, with the R value of the material

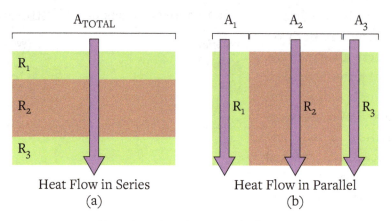

FIGURE 8.6 Combining thermal conductivities of different materials, analogous to springs in parallel or in series.

being analogous to the spring constant. So, the combined R value is simply the sum of all individual values:

$$R_{TOT} = R_1 + R_2 + \cdots + R_n.$$

Now let's combine materials so that heat is being conducted though the assembly in parallel. In this case, heat can flow through multiple paths, each with its own R or U value and area A. The combined U value is the area-weighted average of each individual U value, while the reciprocal of the combined R value is the area-weighted average of the reciprocals of each individual R value:

$$U_{avg} = \left(\frac{A_1}{A_{TOT}}\right)U_1 + \left(\frac{A_2}{A_{TOT}}\right)U_2 + \cdots + \left(\frac{A_n}{A_{TOT}}\right)U_n =$$

$$= \frac{1}{R_{avg}} = \left(\frac{A_1}{A_{TOT}}\right)\frac{1}{R_1} + \left(\frac{A_2}{A_{TOT}}\right)\frac{1}{R_2} + \cdots + \left(\frac{A_n}{A_{TOT}}\right)\frac{1}{R_n}$$

Building envelopes are typically constructed by attaching exterior walls on one side of the structural members and interior walls on the other and then filling the cavities between members with some sort of insulation. The structural members typically have a higher thermal conductance than the insulation. Taken as an assembly, the total thermal conductivity of the wall can be calculated using the earlier equations. Considering each material individually, however, heat flowing through the walls tends to choose the path of least resistance (or highest conductivity), and so heat will flow faster through the structural members. This phenomenon is called **thermal bridging**, because the structural members act as a bridge between the indoor and out-door environments. This is best observed on a winter's day with the help of an infrared camera, as seen in **figure 8.7**.

One goal of architects and structural engineers is to choose materials and assemblies that min-imize thermal bridging while still providing sufficient structural support. This is not just a matter of choosing materials with low thermal conductivities. Again, considering the example of fram-ing members, a material with a relatively low compressive strength will require a larger cross-sectional area, thus increasing the amount of heat flow through the members according to Fourier's law.

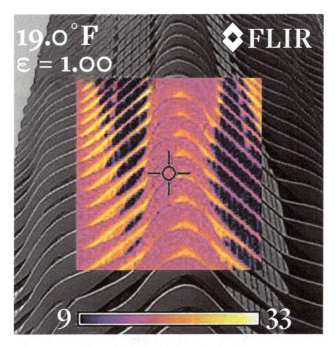

FIGURE 8.7   Thermal bridging of floors in a building.

VI. Recyclability and Recycling Rates

We have discussed sustainability considerations for production and use of civil engineer-ing materials, but it is also important to consider what happens to them after bridges are replaced and buildings are torn down. The primary metric that is used to compare materials is **recyclability**, or whether they can be collected, separated, and made into a new product with comparable performance. In principle, everything *can* be recycled if we are willing to spend the energy, time, and money to separate individual layers, or even molecules, of a product, and then reassemble these individual constituents. But recyclability typically reflects how easy it is to capture the bulk material without too much work.

There are degrees of recyclability. Metals are said to be infinitely recyclable: they can be re-melted into new metal products over and over again. Paper is recyclable, but there is a limit: Each time paper gets recycled, the lengths of its fibers are reduced until they are so small that they can no longer be manufactured into new paper. Materials that are used in a product of lesser quality or simpler function are said to be **downcycled**, for example, plastic bottles that are re-melted into plastic decking rather than back into plastic bottles.

It is also possible for materials to be **upcycled**, for example, by using the plastic bottles as the feedstock for "fleece" textiles.

Recycling a material into a new product means that all of the energy use and emissions that come from extracting, processing, and transporting virgin material can be avoided. In other words, we can save the embodied energy use and emissions of the virgin material. For large volume engineering materials, such as steel, one year's worth of recycling (some 570 million metric tons of scrap globally in 2011) saves approximately 15 EJ (10^{18} J), enough primary energy to power New York City for more than a decade!

A material can be recyclable but not actually get recycled. Look in any public trash can, and you are likely to find many items that could be recycled: paper, plastic bottles, or aluminum cans. The actual degree to which a recyclable material is actually recovered and recycled into new products is called its **recycling rate**. This is the proportion of a material that would otherwise enter the waste stream that is instead recycled. Like embodied energy, the recycling rate is a circumstantial metric—it depends on local circumstances, such as the equipment and infrastructure available for collection and separation, local policies and regulations, prevailing economics, and the availability of end uses that could make use of recycled (or downcycled) material. As mentioned earlier, metals are in principle infinitely recyclable, but for most of the metals in the periodic table, end-of-life recycling rates are well below 10 percent (**figure 8.8**). Iron and steel have one of the highest global recycling rates at 70–90 percent (across a range of estimates), largely because of the fact that ferrous scrap can be recovered and separated easily using magnets.

For metals specifically, **recycled content** is another sustainability metric that can be useful in choosing among materials and suppliers and is one that is commonly considered by rating and certification systems. Recycled content measures the proportion of metal that came from secondary (or recycled) metal versus metal from virgin sources. The recycling rate tends to be correlated with recycled content, but the correlation is not perfect because of the time lag between when a metal is originally produced and when it becomes available for recycling as scrap. **Figure 8.9** shows recycled content for all metals, which can be compared to the results in **figure 8.8**.

VII. Life-Cycle Assessment and Data Resources

In this chapter, you've considered the environmental sustainability of materials from the point of their production, their use, and their end of life, looking at both energy and emissions. That's a lot to juggle! Nevertheless, ideally we would like to evaluate civil engineering materials holistically, in a way that follows what actually happens to them over their entire technological lifetime, from "cradle to grave," considering all the important ways in which

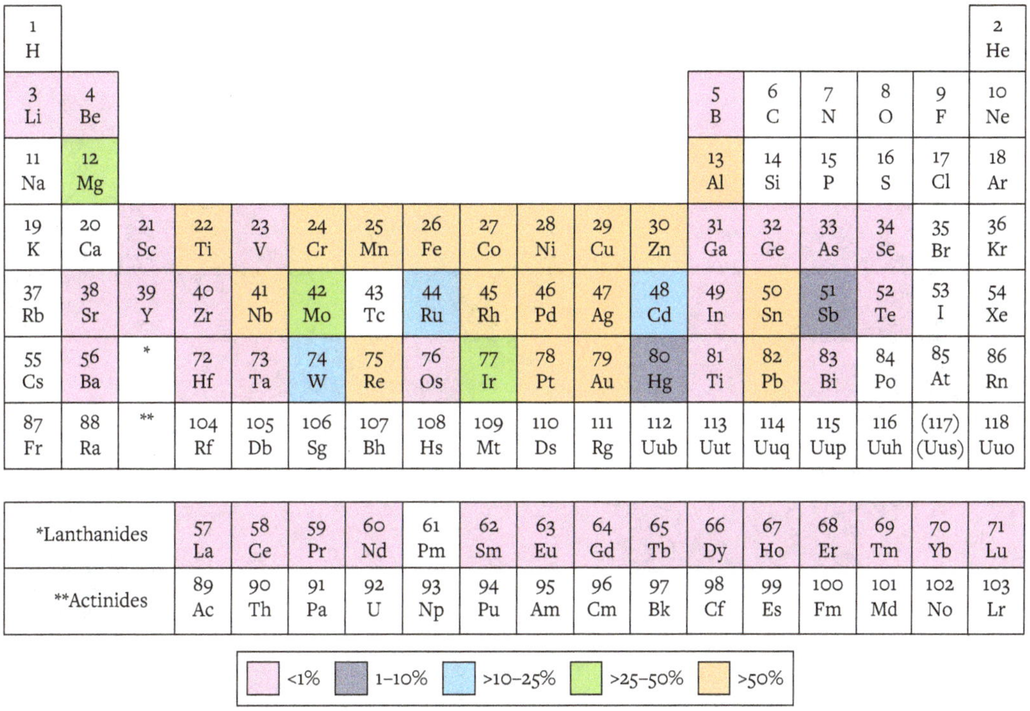

1																	2
1 H																	2 He
3 Li	4 Be											5 B	6 C	7 N	8 O	9 F	10 Ne
11 Na	12 Mg											13 Al	14 Si	15 P	16 S	17 Cl	18 Ar
19 K	20 Ca	21 Sc	22 Ti	23 V	24 Cr	25 Mn	26 Fe	27 Co	28 Ni	29 Cu	30 Zn	31 Ga	32 Ge	33 As	34 Se	35 Br	36 Kr
37 Rb	38 Sr	39 Y	40 Zr	41 Nb	42 Mo	43 Tc	44 Ru	45 Rh	46 Pd	47 Ag	48 Cd	49 In	50 Sn	51 Sb	52 Te	53 I	54 Xe
55 Cs	56 Ba	*	72 Hf	73 Ta	74 W	75 Re	76 Os	77 Ir	78 Pt	79 Au	80 Hg	81 Tl	82 Pb	83 Bi	84 Po	85 At	86 Rn
87 Fr	88 Ra	**	104 Rf	105 Db	106 Sg	107 Bh	108 Hs	109 Mt	110 Ds	111 Rg	112 Uub	113 Uut	114 Uuq	115 Uup	116 Uuh	(117) (Uus)	118 Uuo

*Lanthanides	57 La	58 Ce	59 Pr	60 Nd	61 Pm	62 Sm	63 Eu	64 Gd	65 Tb	66 Dy	67 Ho	68 Er	69 Tm	70 Yb	71 Lu
**Actinides	89 Ac	90 Th	91 Pa	92 U	93 Np	94 Pu	95 Am	96 Cm	97 Bk	98 Cf	99 Es	100 Fm	101 Md	102 No	103 Lr

Legend: <1% | 1–10% | >10–25% | >25–50% | >50%

FIGURE 8.8   End-of-life recycling rates for all metals.[10]

materials can affect the environment and our health. This can be done using a modeling framework called **life-cycle assessment** (LCA).

LCA has its origins in the late 1960s and 1970s when it was created to help companies make decisions about materials for packaging and consumer products, first in the US and then spreading to Europe and Japan. The tool was formally developed in the 1990s and is governed by several International Standard Organization (ISO) standards in the 14000 (environmental management) series. LCA is a generic technology assessment tool, which means that it is flexible and can be applied to just about any physical material or process at any scale; LCA studies have been conducted from the nanoscale all the way up to the entire global economy. It is most commonly applied, however, to individual materials and components/assemblies to help inform product design decisions and consider trade-offs among design alternatives.

LCA compiles resource use and emissions over the entire life cycle of a product and then relates those inputs and emissions to environmental and health impacts. Typically,

---

10   Markus Reuter, Christian Hudson, Antionette van Schaik, Kari Heiskanen, Christina Meskers and Christian Hagelüken, "UNEP Metal Recycling: Opportunities, Limits, Infrastructure, A Report of the Working Group on the Global Metal Flows to the International Resource Panel," 2013, https://www.resourcepanel.org/reports/metal-recycling.

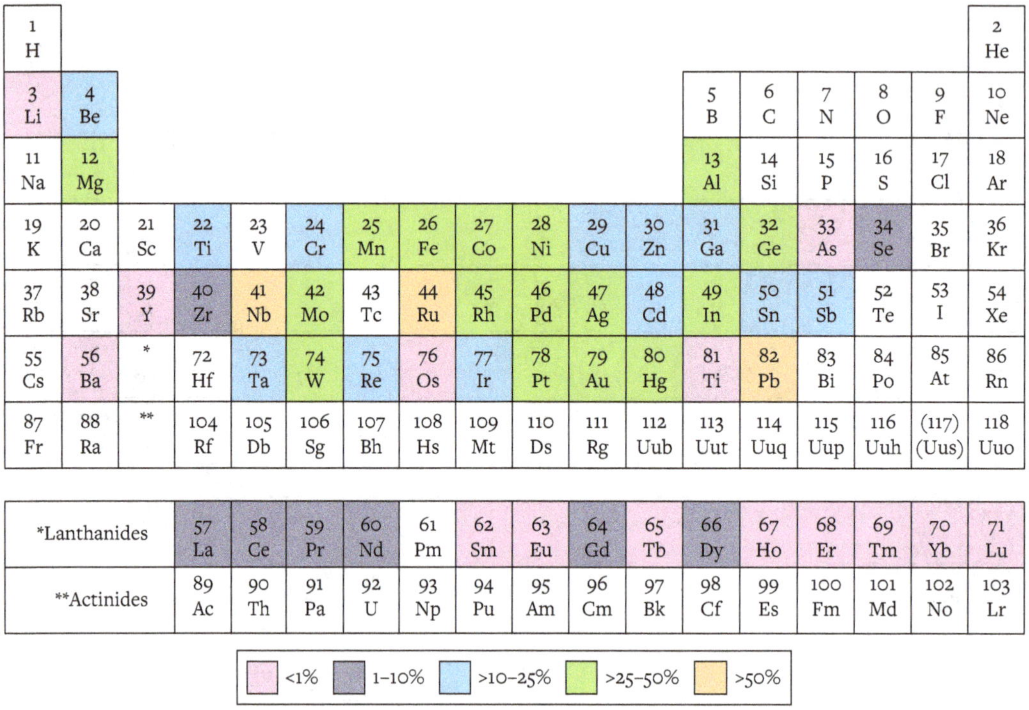

| | <1% | 1–10% | >10–25% | >25–50% | >50% |

FIGURE 8.9   Recycled content percentages for all metals.[11]

several types of impacts are considered, not only energy use and GHG emissions that we have discussed so far, but also other resource inputs such as water, environmental changes such as acid rain and depletion of the stratospheric ozone layer, and human health impacts such as cancer and respiratory disease. By considering multiple impact categories over different life-cycle stages of a product, LCA is ideal for identifying **hotspots**, or individual materials or processes in the life cycle that give rise to a majority of impacts. For example, a hotspot in the life cycle of a laptop computer is the refining of silica to ultra-high purity (99.999999999 percent) silicon for semiconductor components, which requires significant amounts of energy. Different hotspots can occur simultaneously throughout the life cycle, so their identification helps designers balance trade-offs, for example, between making an engineering material that lasts forever but takes a lot of energy to produce and one that requires less energy but might require periodic maintenance or replacement.

Formally, LCA has four phases: (1) goal and scope definition, where the product or system under study is defined and modeling assumptions are documented; (2) life-cycle inventory (LCI), which is a total accounting of resource inputs and emissions over the life cycle;

11   Ibid.

(3) life-cycle impact assessment, which follows how emissions are transported and transformed through the environment and eventually cause physical or biological changes that give rise to negative impacts; and (4) interpretation, where model sensitivity and uncertainty are tested and results are related back to design decisions.

Performing an LCA can be time intensive, gathering emissions data from dozens of suppliers or modeling the effects of thousands of substances in air, water, and soils. Luckily, many LCA tools and databases exist that greatly simplify the job of collecting and calculating LCA results. Here are some of the best free resources for civil engineering applications:

- US LCI database (www.lcacommons.gov): Contains full LCI data (resource inputs and emissions) for dozens of industrial materials and processes specific to the US
- Athena Sustainable Materials Institute (www.athenasmi.org): Packages LCI data on 300-plus building materials into free LCA tools, including impact estimator for buildings and pavement LCA
- NIST Building for Environmental and Economic Sustainability (http://www.nist.gov/el/economics/BEESSoftware.cfm): Includes an LCI database and online tool for assembling and comparing different building materials
- World Steel Association (http://www.worldsteel.org/steel-by-topic/life-cycle-thinking.html)
- National Ready-Mix Concrete Association (http://www.nrmca.org/sustainability/epdprogram/LCA.asp)

VIII. Concluding Remarks

In many situations, specifying a material is straightforward: Local building codes may restrict you to a single choice, or you may simply rely on industry convention or the practices of your particular firm. When there are materials choices to be made, however, it can be difficult to know *how* to make a decision. This chapter has discussed some of the more important environmental considerations for civil engineering materials, but how do you know which are most important, how to balance against cost and other factors, and which are the appropriate targets? Here are a few points from the chapter to help you organize your decisions:

- First, remember that all comparisons should be made on the basis of a functional unit; you should always compare materials based on equivalent performance, rather than equivalent mass. Visualization tools, such as Ashby plots, can be used to consider functional characteristics of materials against other sustainability metrics.

- Take a life-cycle approach. Civil engineers design and build projects that must perform reliably over long periods of time. Consider both the environmental implications of materials used in original construction and how those materials will affect energy use or emissions over the use phase of the project, particularly if materials require maintenance or replacement.

- Sustainability does not just mean low energy and GHG emissions. These days, businesses and consumers care about numerous environmental issues all at the same time. LCA tools will help you compare products across multiple types of environmental and health impacts. Consider especially how your materials may expose people to harmful substances—for example, off-gassing of volatile organic compounds or leaching of metals—and make sure the materials you use have undergone appropriate testing.

- There are many win-win situations. Fly ash can improve concrete performance for certain applications; recycled materials often save costs and perform just as well as virgin materials; specifying low-emission materials in buildings may improve occupant comfort and productivity. Understanding the potential material co-benefits for your particular project will help you justify specifying an environmentally preferable material with which others on your team may be less familiar.

- Leverage the growing market for environmentally preferable materials. More and more civil engineering projects are being built to green certification standards, such as Leadership in Energy and Environmental Design for Buildings and Envision™ for infrastructure. Both of these standards include material sustainability considerations, such as recycled content and use of local materials, which means that firms are asking more questions and making more requests about material options, while suppliers are responding by developing product lines with desired environmental attributes. The market for environmentally preferable materials is evolving quickly, so a bit of research may uncover new products and options that did not exist just a couple years ago.

All materials incur environmental impacts of some sort. As the adage goes, "There is no such thing as a free lunch." In the future, we may be able to build structures with materials that clean the air and are net producers of energy and water (and there is much exciting research in science and engineering in this area!), but don't worry if you can't get to zero embodied energy or emissions or minimize all types of environmental impacts at once. Trade-offs are nearly universal; it is your job to understand which sustainability considerations are most important for your project and to prioritize these when choosing among material options.

IX. Problems

1. Considering the "triple bottom line" of sustainability, give two examples of sustainability metrics for each of the three pillars of economic, environmental, and social sustainability, as related to civil engineering materials.

2. What does embodied energy measure? How about embodied carbon or embodied GHG emissions?

3. Aluminum production is an electricity-intensive industrial process, but over the past 15 years, the average electricity used per metric ton of primary aluminum has declined from 15,400 kWh/ton to 14,300 kWh/ton of final energy.[12] Suppose that the electricity is derived solely from natural gas-fired power plants and assume the same conversion efficiencies listed in Example 8.1. What is the primary energy savings associated with this decrease in electricity requirements, measured in kWh/ton?

4. The article "The Growth of Urban Building Stock: Unintended Lock-in and Embedded Environmental Effects"[13] gives the following table of materials for a single-family detached home in California:

Material	Quantity (metric tons)
Aggregate	33.5
Aluminum	0.3
Asphalt shingles	4.6
Concrete (20 MPa)	47.6
Fiberglass batt insulation	0.4
Plywood	3.8
Steel nails	0.3
Steel sheet	0.5
Stainless steel hardware	0.1
Timber, softwood	12.3

Use the embodied energy and embodied carbon values in table 8.1 to determine the embodied energy and carbon for the entire house. Assume virgin materials.

5. Many US States now allow for the use of fly ash as a substitute for cement in concrete. Suppose the house project in question 4 uses concrete where 35 percent of the cement has been substituted with fly ash. If the cement specified is 15 percent of

---

12   "Aluminum," IEA, https://www.iea.org/tcep/industry/aluminium/

13   Janet L. Reyna and Mikhail V. Chester, "The Growth of Urban Building Stock: Unintended Lock—in and Embedded Environmental Effects," *Journal of Industrial Ecology* 19, no. 4, (2014), 524–537.

the concrete mix, what will be the embodied energy and embodied carbon savings of this design choice?

6. Using the Ashby plot in figure 8.4, identify the material that has the highest ratio of strength to embodied energy in the units provided.

7. Portland cement is approximately 65% calcium oxide by mass. Use the chemical equation for thermal decomposition for calcium carbonate to calculate the carbon dioxide associated with the calcium oxide required for 1 kilogram of cement. The relevant molar weights are 100.1 g/mol for $CaCO_3$ and 56.1 for CaO. Why is this number less than the embodied carbon value for cement shown in table 8.1?

8. Choose one of the sustainable production strategies described in section IV. Using industry or government reports or published literature, write one paragraph describing recent trends and/or technological developments.

9. Stainless steel is resistant to corrosion and therefore has a longer material lifetime in use than untreated carbon steel. On the other hand, stainless steel is more energy and carbon intensive to produce than carbon steel. If an exposed steel member is expected to last 60 years, how much longer would a stainless steel member need to last in order justify its use from an embodied carbon point of view? Use the values in table 8.1.

10. Which has a higher R value of thermal resistance: a double-paned window with 2 panes of one-fourth-of-an-inch glass and a half-inch air gap, or 1 inch of rockwool insulation?

11. What is thermal bridging, and what is one way to prevent it?

12. Electrochemical batteries using chemistries such as lithium-nickel-manganese-cobalt oxide (NMC) enable electric vehicles and renewable energy storage. As more NMC batteries get used in products, more will come out of use when products reach end of life. Using the information on elemental recycling in figure 8.8, which elements are most likely to be recovered from NMC batteries, using today's recycling techniques?

13. Detailed next are two alternative foundation designs for a bridge pier. Both designs are equal in their ability to support applied loads and meet other performance criteria (i.e., they are functionally equivalent). The cost for each design is provided, but the environmental impacts have not yet been quantified. Using embodied energy as an indicator of environmental impact, make a recommendation as to which foundation design alternative is the most sustainable option. In addition to the materials, consider all aspects of the construction (materials production, materials and waste

transportation, equipment operations on site) when estimating the total embodied energy in each complete foundation alternative. Assume all steel is virgin steel.

Useful Information:

- Unit weight of steel = 490 pounds per cubic foot
- Unit weight of reinforced concrete = 150 pounds per cubic foot
- Liquid diesel fuel has an embodied energy of 43.0 MJ/L[14], including its energy content at the lower heating value (LHV) and the upstream processing energy.
- The fuel economy of a typical heavy-duty truck is 6 miles/gallon, and all heavy-duty trucks are fueled with diesel (Note: 1 US gallon = 3.78 liters).
- A typical concrete truck carries 10 cubic yards of concrete.
- A typical dump truck for waste disposal carries 20 cubic yards of material.
- Every truck makes a *round trip* for every load it delivers to or takes from the site.

The two design alternatives are as follows:

Design 1:

- The bridge pier is supported by 20 piles, which are 40-feet-long driven steel H-piles, with engineered steel section HP 12 × 84.
- The steel supplier is located 150 miles from the site.
- The steel sections must be transported on a flatbed trucks, four piles per truck.
- The machinery involved in pile driving will consume a total of 500 gallons of diesel fuel to drive the 20 piles (this does not include transportation fuel to deliver the piles to the site).
- Assume no waste transportation for this design.
- Cost of this design: $220,000

Design 2:

- The bridge pier is supported by four drilled shafts, 50 feet long and 3 feet diameter (drilled shafts are cast in place, reinforced concrete piles).
- The concrete compressive strength is 4,000 psi, and it is steel reinforced.
- The concrete batch plant is located 15 miles from the site.

14   Craig M. Shillaber, James K. Mitchell and Joseph E. Dove, "Energy and Carbon Assessment of Ground Improvement Works. II: Working Model and Example." *Journal of Geotechnical and Geoenvironmental Engineering*, 142, no. 3 (2016).

- The steel rebar supplier is located 150 miles from the site and is delivered on one heavy duty truck.
- This design involves disposal of the excavated soil from shaft drilling as waste. The bulking factor for excavated soil is 1.3. (Note: The volume of excavated soil increases over its volume in-situ. In this case, the bulking factor means the volume of excavated soil is 1.3 times its volume in the ground).
- The landfill for disposal of excavated soil is located 10 miles from the site.
- The machinery involved in shaft drilling will consume a total of 700 gallons of diesel fuel to construct the four shafts (this does not include transportation fuel to deliver materials to the site or remove waste from the site).
- Cost of this design: $250,000

Solution hints:

- Determine total quantities of construction materials and waste for each alternative.
- Determine total quantity of fuel for each design alternative (for material and waste transportation and site equipment).
- Use quantities to determine total embodied energy for each design (embodied energy values per unit for construction materials are in table 8.1).
- Use given information (cost) and estimated embodied energy to decide which design is best/most sustainable.

CPSIA information can be obtained
at www.ICGtesting.com
Printed in the USA
LVHW061944100921
697565LV00002B/13